48, 65, 147
49

Other titles published in collaboration between
ELLIS HORWOOD LIMITED and
THE WATER RESEARCH CENTRE

SEWAGE SLUDGE STABILISATION AND DISINFECTION
A. M. BRUCE, WRC
BULKING OF ACTIVATED SLUDGE: Preventative and Remedial Methods
B. CHAMBERS and E.J. TOMLINSON, WRC
STABILISATION, DISINFECTION AND ODOUR CONTROL IN SEWAGE SLUDGE TREATMENT An Annotated Bibliography covering the period 1950-1983
Compiled by Mrs E.S. CONNOR and edited by E.S. CONNOR and A.M. BRUCE, WRC
BIOLOGICAL FLUIDISED BED TREATMENT OF WATER AND WASTEWATER
Edited by P.F. COOPER, WRC and B. ATKINSON, UMIST
WATER RESEARCH TOPICS Volume 1
Edited by I.M. LAMONT, WRC
NEW PROCESSES OF WASTE WATER TREATMENT AND RECOVERY
Edited by G. MATTOCK
RIVER POLLUTION CONTROL
Edited by M.J. STIFF, WRC
AQUALINE THESAURUS
Compiled by J.G. SMITH and edited by P.J. RUSSELL, WRC

ENVIRONMENTAL PROTECTION:
Standards, Compliance and Costs

Editor:

T. J. LACK, B.Sc.(Special Hons.), Ph.D.
Head of Marine Toxicity
and Co-ordinator, Sea Outfalls Research
Water Research Centre
Marlow, Buckinghamshire

Published for the
WATER RESEARCH CENTRE
by

ELLIS HORWOOD LIMITED
Publishers · Chichester

First published in 1984 by
ELLIS HORWOOD LIMITED
Market Cross House, Cooper Street, Chichester, West Sussex, PO19 1EB, England

The publisher's colophon is reproduced from James Gillison's drawing of the ancient Market Cross, Chichester.

Distributors:
Australia, New Zealand, South-east Asia:
Jacaranda-Wiley Ltd., Jacaranda Press,
JOHN WILEY & SONS INC.,
G.P.O. Box 859, Brisbane, Queensland 40001, Australia

Canada:
JOHN WILEY & SONS CANADA LIMITED
22 Worcester Road, Rexdale, Ontario, Canada.

Europe, Africa:
JOHN WILEY & SONS LIMITED
Baffins Lane, Chichester, West Sussex, England.

North and South America and the rest of the world:
Halsted Press: a division of
JOHN WILEY & SONS
605 Third Avenue, New York, N.Y. 10016, U.S.A.

© 1984 Water Research Centre/Ellis Horwood Limited

British Library Cataloguing in Publication Data
Environmental protection.
1. Water — Pollution
I. Lack, T. J. II. Water Research Centre
363.7'394 TD420

ISBN 0-85312-740-9 (Ellis Horwood Limited)
ISBN 0-470-20095-2 (Halsted Press)

Typeset in Press Roman by Ellis Horwood Limited.
Printed in Great Britain by R.J. Acford, Chichester.

COPYRIGHT NOTICE —
All Rights Reserved. No part of this publication may be reproduced, stored in a retrieval system, or transmitted, in any form or by any means, electronic, mechanical, photocopying, recording or otherwise, without the permission of Ellis Horwood Limited, Market Cross House, Cooper Street, Chichester, West Sussex, England.

Table of Contents

Editor's Preface. 9

Directory of Contributors. 11

Keynote Address – Environmental protection: standards, compliance,
    and costs. 14

PART 1 – THE NATURE AND ASSOCIATED RISKS OF WATER POLLUTION

Chapter 1 – The origins and pathways of aquatic pollutants
    N. J. King, P. R. Hinchcliffe and J. L. Vosser, Department
    of the Environment 22

Chapter 2 – The nature and concentration of pollutants in water
    M. Fielding and D. T. E. Hunt, Water Research Centre. 33

Chapter 3 – Pollution – the risk in perspective
    Sir Frederick Warner, FRS, Cremer and Warner 55

Discussion 69

PART 2 – ASSESSMENT OF EFFECTS OF POLLUTION

Chapter 4 – The role of toxicology
    J. K. Fawell, Water Research Centre. 72

Chapter 5 – Epidemiology and water quality
    Patricia Fraser, London School of Hygiene and Tropical
    Medicine. 85

Chapter 6 – Ecotoxicology and environmental quality
    W. H. Könemann, Ministry of Housing, Physical Planning
    and Environment, The Netherlands 94

Discussion 104

PART 3 – PROTECTION AND CONTROL

Chapter 7 – Standards and limits for water quality control
R. F. Packham, Water Research Centre 108

Chapter 8 – Monitoring compliance with standards
J. C. Ellis and D. G. Miller, Water Research Centre. 123

Discussion 140

Chapter 9 – The role and activities of the Oslo and Paris Commissions
F. Bjerre and P. A. Hayward, Oslo and Paris Commissions. . . . 142

Chapter 10 – European Community policy on environmental management
A. J. Fairclough, Commission of the European Communities, Belgium. 158

Chapter 11 – The approach of the Federal German Republic – anticipating environmental effects
Professor J. Salzwedel, The Council of Experts on Environmental Matters, Federal German Republic 169

Chapter 12 – The UK approach – environmental quality standards
J. A. L. Gunn, Department of the Environment 181

Discussion 189

PART 4 – OPERATIONAL ASPECTS

Chapter 13 – Pollution control in practice
B. Alexander and E. Harper, North West Water Authority. . . . 198

Chapter 14 – Spillages – an operational approach
J. A. Young and R. J. Whitaker, Wessex Water Authority 209

Chapter 15 – Contingency planning for chemical accidents
A. Gilad and J. I. Waddington, World Health Organization Regional Office for Europe, Denmark. 219

Discussion 234

Chapter 16 – Detection of pollution at drinking water intakes
G. P. Evans and D. Johnson, Water Research Centre 239

Discussion 255

PART 5 – ENVIRONMENTAL MANGEMENT – THE PRIORITIES

Chapter 17 – Costs of environmental quality to the water industry
W. R. Harper, Thames Water Authority, and
J. Moss, Water Research Centre 258

Table of Contents

Chapter 18 – The impact on manufacturing industry
T. Farquhar, Albright and Wilson Limited 267

Chapter 19 – Environmental protection – what is it worth?
Geoffrey Lean, *The Observer* 273

Discussion 281

Chapter 20 – Protection of the aquatic environment – research needs
S. C. Warren, Water Research Centre 283

Chapter 21 – Environmental protection – what can the nation afford?
Lord Sherfield, House of Lords Select Committee on
Science and Technology. 295

Discussion 300

Summing Up 304

Appendix I – Support paper
The perception of risk
P. Powell and R. F. Lacey, Water Research Centre 309

List of Delegates 315

Index 325

Editor's Preface

This book has its origins in the Water Research Centre Conference entitled Environmental Protection: Standards Compliance and Costs, held at Keele University, Staffordshire, in October 1983. The conference was arranged to provide a forum for a critical examination of the scientific basis and operational approaches to pollution control in the water industry. The conference was opened by Mr J. G. Bellak, Chairman of Severn-Trent Water and former Managing Director of the Royal Doulton Tableware Group. Mr Bellak is a keen ornithologist and conservationist and has had many years' experience in negotiations in Brussels on standards for heavy metal emissions by the pottery industry, and his point that protection of the environment is not just a technical matter but one of growing social and political importance set the scene for all those who attended.

The book will interest those who are involved in the legislative, scientific, and operational areas of freshwater and marine environmental protection.

The introductory Keynote Address is written by Lord Ashby the eminent scholar, Chairman of the Working Party on Pollution Control at the United Nations Conference on the Environment in Stockholm, 1972, and Chairman of the Royal Commission on Environmental Pollution from 1970 to 1973.

The authors represent broad strands of environmental interests including United Kingdom, European government, and European Community policies. In the book special attention is paid to the criteria used in the judgement of compliance with standards, the balancing of costs and risks and the priorities for pollution control measures. Operational approaches to pollution control and protection of the environment are described by reference to specific case studies and contingency planning. The reader will also gain an insight into the uses of toxicology and epidemiological studies to assess the damage that man is doing to himself and his planet.

The final contribution, a summing up of the conference papers and discussion, is written by Professor R. W. Edwards, Head of the Department of Applied

Biology, University of Wales Institute of Science and Technology, and Deputy Chairman of Welsh Water.

A great deal of lively discussion occurred during the conference and, as it has been necessary to condense the discussions into summaries of the responses of authors to comments and questions, the Water Research Centre thanks all those participants who contributed to the discussions but who cannot be named.

The Water Research Centre would also like to thank the authors for their contributions to the Conference, now chapters in this book, and the session chairmen. Thanks are also due to our guest speaker John Maddox, Editor of *Nature*, who gave a thoughtful and humorous speech after the conference dinner.

Acknowledgement is made to Her Majesty's Stationery Office, who reserve the copyright on Chapters 1 and 12.

T. J. LACK
Medmenham, January 1984

Directory of Contributors

		Chapter
Alexander, B.	Director of Planning, North West Water, Dawson House, Great Sankey, Warrington, Cheshire	13
Bjerre, F.	Deputy Secretary, Oslo and Paris Commissions, New Court, 48 Carey Street, London	9
Ellis, J. C.	Head, Statistics, WRC Environment, Henley Road, Medmenham, PO Box 16, Marlow, Bucks. SL7 2HD	8
Evans, Dr G. P.	Head, Intake Protection, WRC Environment, Henley Road, Medmenham, PO Box 16, Marlow, Bucks. SL7 2HD	16
Fairclough, A. J.	Director for the Environment, Commission of the European Communities, Rue de la Loi 200, B-1049, Brussels, Belgium	10
Farquhar, Dr J. T.	Group Environmental and Protection Manager, Albright and Wilson Limited, 1 Knightsbridge Green, London	18
Fawell, J. K.	Head, Toxicology, WRC Environment, Henley Road, Medmenham, PO Box 16, Marlow, Bucks. SL7 2HD	4
Fielding, M.	Head, Organics, WRC Environment, Henley Road, Medmenham, PO Box 16, Marlow, Bucks. SL7 2HD	2
Fraser, Dr Patricia	Department of Medical Statistics and Epidemiology, London School of Hygiene and Tropical Medicine, Keppel Street, London	5

Directory of Contributors

Gilad, Dr A.	Consultant, Environmental Systems Management, World Health Organization Regional Office for Europe, 8 Scherfigsvej, 2100 Copenhagen, Denmark	15
Harper, E.	Chief Scientific Adviser, North West Water, Dawson House, Great Sankey, Warrington, Cheshire	13
Harper, W. R.	Director of Finance, Thames Water, New River Head, 177 Rosebery Avenue, London	17
Hayward, P. A.	Oslo and Paris Commissions, New Court, 48 Carey Street, London	9
Hinchcliffe, P. R.	Department of the Environment, Romney House, 43 Marsham Street, London	1
Hunt, Dr D. T. E.	Head, Analysis, WRC Environment, Henley Road, Medmenham, PO Box 16, Marlow, Bucks. SL7 2HD	2
Johnson, D.	Biologist, Intake Protection, WRC Environment, Henley Road, Medmenham, PO Box 16, Marlow, Bucks. SL7 2HD	16
King, Dr N. J.	Head, Toxic Substances Division, Department of the Environment, Room A3.335, Romney House, 43 Marsham Street, London	1
Könemann, Dr W. H.	Ministry of Housing, Physical Planning and Environment, Postbus 439, 2260 AK Leidschendam, The Netherlands	6
Lacey, R. F.	Head, Epidemiology, WRC Environment, Henley Road, Medmenham, PO Box 16, Marlow, Bucks. SL7 2HD	Appendix I
Lean, G.	Environment Correspondent, *The Observer*, 8 St Andrews Hill, London	19
Miller, Dr D. G.	Assistant Director, Environmental Contamination, WRC Environment, Henley Road, Medmenham, PO Box 16, Marlow, Bucks. SL7 2HD	8
Moss, Dr J.	Manager, Sludge Disposal and Operational Studies, WRC Environment, Henley Road, Medmenham, PO Box 16, Marlow, Bucks. SL7 2HD	17

Directory of Contributors

Packham, Dr R. F.	Assistant Director, Water Contamination, WRC Environment, Henley Road, Medmenham, PO Box 16, Marlow, Bucks. SL7 2HD	7
Powell, P.	Epidemiology, WRC Environment, Henley Road, Medmenham, PO Box 16, Marlow, Bucks. SL7 2HD	Appendix I
Salzwedel, Professor J.	Chairman, The Council of Experts on Environmental Matters, Lennestrasse 35, D-5300 Bonn 1, Federal Republic of Germany	11
Sherfield, The Lord	House of Lords Select Committee on Science and Technology, Westminster, London	21
Vosser, J. L.	Department of the Environment, Romney House, 43 Marsham Street, London	1
Waddington, J. I.	World Health Organization Regional Office for Europe, 8 Scherfigsvej, 2100 Copenhagen, Denmark	15
Warner, FRS, Sir Frederick	Emeritus Partner, Cremer and Warner, Consulting Engineers, 22 Beach Road, Hartford, Northwich, Cheshire	3
Warren, Dr S. C.	Director, WRC Environment, Henley Road, Medmenham, PO Box 16, Marlow, Bucks. SL7 2HD	20
Whitaker, R. J.	Regional Emergency and Security Officer, Wessex Water, Wessex House, Passage Street, Bristol, Avon	14

Keynote Address – Environmental protection: standards, compliance and costs

ERIC ASHBY

For a long time I gazed at the title of this conference and the titles of the papers to be given. It seems to me (I may be wrong) that nearly all the papers are about *how* to protect the environment, not why to protect it. And this started off some speculations. For centuries laws have been passed and regulations made to protect people against hazards from the environment: floods, contamination of water supplies, and so on. But it is only recently that laws have been passed with what seems to be the sole purpose of protecting the environment from hazards caused by people. Under the Wildlife and Countryside Act 1981 you are forbidden, under pain of £1000 fine, to eject bats that have squatted in your garden shed or garage. The same Act protects wetlands used by migrating birds. Throughout the European Community there is expected to be compliance with a directive to protect freshwater fish, not for the benefit of fishermen but solely for the benefit of the fish. Near the City of Houston in Texas developers have been refused permission to build new suburbs on thousands of acres, in order to preserve the habitat of the rare Houston toad, a creature distinguishable from ordinary toads mainly by its mating call in the spring.

The plain fact is that we have not got any clear idea why we pass these laws to protect natural objects other than ourselves. The common assumption, that all laws are passed out of self-interest, does not fit some of the laws taken for 'environmental protection': the title of this conference. This may seem to be an academic quibble. I suggest to you that it's nothing of the kind. To turn to the subtitle of this conference: 'standards, compliance, and costs', I believe that the point of reconciliation between these three, often conflicting, elements does depend very greatly upon the assumption made about *why* the environment should be protected. Where self-interest is obviously at stake, as in the standards to be set for drinking water, people are willing to comply and willing to pay the cost. But what about compliance with the EEC directive to protect freshwater fish?

The National Water Council in its evidence to a House of Lords Select Committee estimated that the cost of compliance in Britain would be of the order of some thousand million pounds and that in many rivers fish are flourishing even though the rivers fail to comply with some of the standards set by the directive. The fish evidently hadn't read the directive. Here is a case — and there are many others — where the point of reconciliation between standards, compliance, and costs, depends on an elusive ethical query, namely should you regard fish, as the law already does seem to regard bats and other listed animals and plants in Britain, and the Houston toad in Texas, as though they had 'rights'?

The problem is further compounded by the fact that each of the three words in the subtitle to this conference is the province of a different expert: the scientist recómmends standards, e.g. in symbols of micrograms per litre, the lawyer and administrator draw up terms for compliance in cumbersome legalistic prose, the economist works out the costs and benefits in symbols of money. It is, in the end, the politician who has to integrate these symbols into a decision, i.e. to make what your programme calls the 'realistic appraisal'. The expertise he brings to this is 'hunch', though it is dignified by the label 'political judgement'.

I shall have more to say about hunch later on. Let us first consider the contributions made to the political decision by the advisers who draw up standards and costs. The scientist can say *how* clean a river is, and even how clean it needs to be to fulfil stated requirements: e.g. to be safe to drink, to swim in, to be safe for salmon, perch or just duckweed and tadpoles. What he cannot say is whether the river is clean *enough* unless he knows *why* it is required to be clean, by whatever authority is responsible. That is a decision not within the scientist's competence. Something similar can be said about the economist's advice. He can say what it will cost to achieve a stated level of cleanliness. He can even suggest (though I personally have no faith in his calculations) the quality-standard below which the marginal damage done by pollution would exceed the marginal cost of abatement, and above which the marginal cost of abatement would exceed the marginal cost of damage done if you didn't abate it. I have no faith in these calculations because you can't make a credible estimate of marginal damage: the shadow pricing of some amenities may be intellectually acceptable but it's not morally or politically acceptable. Where the economist's advice is useful is on the so called 'opportunity costs', i.e. the other desirable choices you'll not be able to make if you spend more resources on cleaning the river. But the economist deals only with the logic of choice, not with the ethics of choice. So he's not competent to say whether enough is being spent on abatement unless he is told why pollution has to be abated: is it just to serve the interests of people who use the river? If so, is it just their material interests (which includes their rights to discharge pollutants into it)? Or are aesthetic interests to be included? And if the purpose is not solely self-interest, e.g. if it is the sort of motive that inspires people to support the Save the Whales campaign though most of them never expect to see a whale, how are these altruistic motives to be costed? What price the Houston toad, the Teesdale sandwort (a little plant that came near to blocking ICI's project in the 1960s for a dam to provide water for its ammonia plant on Teesside)? Unless you have some idea why you want to spend money on these objects, how can you decide how much to spend? Environmental protection now costs the governments of industrialised countries a great deal of money; their decisions rest on some tacit and implicit environmental ethic. If this so called keynote address has any message at all it is to say that it is high time this environmental ethic was made more explicit [1].

Which brings me to the third word in the subtitle of this conference: compliance. In a pluralistic democracy people expect to be told not just how but why they should comply with regulations made by their elected representatives. Indeed, politicians shrink from making regulations unless they are what is called 'socially acceptable'. Lest you think I am drifting into a philosophic stratosphere let me illustrate this by a very down-to-earth example: the present controversy over the application of the polluter-pays-principle (PPP) to discharges from industrial plants into rivers.

It is risky to make any simple statement about PPP – the concept is in fact very confusing and means different things to different people – but here is a summary of one part of the concept to illustrate the point I want to make. Industrialists in Britain (speaking through such institutions as the CBI and CIA) are willing to pay the cost of purifying effluents to meet the consent-conditions imposed by regional water authorities (RWAs) if they discharge direct into the river. This they regard as the cost of complaince with the law which is made (to quote one definition) 'to ensure that the environment is in an acceptable state' – whatever that means. Also they are willing to pay the charges if they put their effluent into a public sewer. This they regard as payment for services rendered. But industrialists argue that any pollutant discharged within the consent conditions, called 'residual pollution' (though nationwide is amounts to immense quantities), should cost them nothing. The assumption is that residual pollution does not impose social costs and therefore it is not 'acceptable' to industrialists to be made to pay for it. If residual pollution is deemed to be harmful, then it is up to the RWA to stiffen consents so that the river is in 'an acceptable state', i.e. is clean enough.

It is these words, 'acceptable' and 'enough' that are so elusive because we have no clearly defined environmental ethic. In Britain the words cover a very flexible policy. We classify rivers into classes I to IV and do not aspire to raise all rivers into class I. The RWA makes for each river what are political and even ethical decisions about what is acceptable and 'clean enough'.

Other people, in Britain and elsewhere, take a totally different view of the meaning of PPP. They do not object to the controlled discharge of wastes into rivers but they believe that water (and air, too, and land) should not be free goods to be exploited gratis. Everything discharged into the environment, they say, involves some social cost. Accordingly those who hold this view believe that the industrialist should be obliged to pay for his residual pollution even if it is within consent conditions. To put it another way, he should have to buy pollution rights in air, water, or on land if he wants to dump his wastes there. Payment for all pollution (as I expect you'll hear tomorrow morning) is the principle on which Germany, France, and the Netherlands base their policies for river management.

There is an elegant economic argument to support this principle: to control pollution by market forces (even to auction the right to pollute) would be the most cost-effective way to do it; it would optimise the behaviour of each individual polluter. I have no time to comment on this except to say that there is no convincing evidence yet that this interpretation of PPP in some EEC countries *is* proving more cost-effective. (You will find a very full discussion of this in the report of the House of Lords Select Committee on the European Communities, entitled *The Polluter Pays Principle*, published recently [2]. There is also a common sense argument to support the principle. It is that to subject polluters to a continuing pressure to abate their pollution by charging them for all wastes deposited outside their own premises will ensure that the environment will become progressively cleaner, to match the progressively rising standards of what the public regard as acceptable. By contrast, our system in Britain is regarded by some people as a formula for complacency. The consent sets the level of what is clean enough − 'enough' in this context is a purely pragmatic decision − and there may be no incentive to make the river any cleaner.

I believe that these divergent views about PPP rest on different tacit ethical assumptions about why the environment should be cleaned at all beyond levels necessary to ensure public safety. The ultimate aim of environmental policy in the European Community is to *minimise*, not to *optimise* pollution. The Congress of the USA apparently had a similar aim when it passed the Federal Water Pollution Control Act Amendments in 1972, declaring that the discharge of pollutants into the nation's rivers should be eliminated by 1985 and most rivers should be 'fishable and swimmable' by then (an aim which has now been postponed into a remote future).

British policy for river management holds that 'clean enough' is what the public will 'tolerate' (a word used by Chemical Industries Association); but − as Lord Sherfield will remind you in his paper on Thursday − 'the standards of one generation will not suffice for the next'. The Community policy for river management holds that there is no finality about 'enough'; pollution, like crime, should be minimised, not optimised.

I think those responsible for environmental policy in the water industry are going to run into a further difficulty arising from our hazy ideas about why we legislate to protect the environment from hazards caused by man. In its deplorably belated response to the fifth report of the Royal Commission on

Environmental Pollution the present government concurred in an important recommendation from the Commission, namely that policy for the disposal of wastes should be guided by choice of the best practicable environmental option — a refinement of the very successful tradition in the Alkali Inspectorate, to require industry to use the best practicable means for abating air pollution in registered processes. But a policy based on best practicable means is not consistent with one based on the consent system in rivers, for there is no doubt that discharges into rivers could be much cleaner than they are if RWAs insisted on the use of best practicable means instead of insisting only on compliance with consents. It seems now that the time must come (I hope it comes soon) when there is co-ordination between RWAs, the Health & Safety Executive, and local authorities over the whole strategy of waste disposal, to decide whether the waste shall go into water, or the air, or on land.

In all this the role of the scientist and the economist become secondary. Indeed some very major decisions are being made at this very time which are inconsistent with the advice scientists and economists give to governments. Let me give you an example from a pollutant that underlines the need for closer co-operation between policies for air and for water: the control of emissions of sulphur dioxide, and the hazard of acid rain, especially on fisheries. It is not surprising that the Central Electricity Generating Board (CEGB) is opposed to any regulation that would oblige them to cut down emissions of sulphur dioxide from power stations. But there is now a wave of sentiment in Europe and North America hostile to acid rain. There is (in my view) little evidence that sulphur dioxide emissions are any longer doing harm in Britain: of course at one time they did great harm, but mean ground level concentrations have been reduced to an average of less than $80 \mu g/m^3$; our soils are well enough buffered not to be damaged by rain at pH 4.5. But we do export sulphur dioxide to Scandinavian countries, where acid rain does affect some lakes and kills fish. Should we be good neighbours and reduce our SO_2 emissions? The scientific evidence is indecisive: it may be the concentrations of oxidants in the air, necessary to convert SO_2 into sulphuric acid, that determine the amount of acidity in the rainfall; if so, to halve the emissions of SO_2 would not halve the acid rain — it might have little effect on it. In any case, even if Britain eliminated sulphur dioxide altogether from the chimney stacks of power stations, it would have little effect in Scandinavia unless half a dozen other nations did the same thing. And the economist's evidence — in this case incontrovertibly decisive — is that the cost of even halving the amount of SO_2 would be very high indeed.

Canada faces an analogous problem. Over 70 per cent of the SO_2 deposited in Ontario doesn't come from Canada at all: it comes from the USA. A reduction in sulphur emissions in Ontario would have only a trivial effect on the acidity of the rain falling on Ontario's lakes. And yet a regulation has already been passed that Ontario Hydro (the Ontario equivalent of the CEGB) must reduce its SO_2 emissions by 50 per cent by 1990.

There's no doubt that Britain will be obliged to reduce sulphur emissions, just as Canada has been obliged to, because social values voiced through pressure groups and amplified by the mass media and shown in technicolour by investigative journalists on television, now demand that sulphur dioxide, like lead, shall be blacklisted. Whitehall is already on record as saying that the 'high chimney' policy for exporting our sulphur dioxide is 'obsolete'. Politicians will have to respect the current norm of cleanliness which rejects sulphur dioxide in air, as our grandparents rejected untreated sewage in drinking water.

Which brings me back to 'hunch'. The history of environmental protection is a record of piecemeal responses to specific episodes: the River Pollution Acts to outbreaks of cholera and typhoid; the Clean Air Acts to the 1952 London smog; the pesticide control policy to the disclosures in *Silent Spring*; the Deposit of Poisonous Waste Act to the discovery of cyanide drums dumped at Nuneaton. It gives the impression that decision makers in Whitehall and Westminster are no more than seismographs recording public upheavals. But it's more subtle than that. It is the legislator who has to make what your programme calls the 'realistic appraisal' and the evidence of public opinion, even if sometimes misguided or ignorant of the facts, is likely to weigh more heavily than the evidence of scientists or economists. And for a good reason: the gap between the rational advice of experts and its incorporation into statutes cannot be closed unless the statute makes explicit what is already widely implicit in the values held by the people: 'the opinion of the governed is the real foundation of all government'. Indeed some statutes have to be made just to codify what have become fresh values even though some experts are unconvinced of their necessity. How these fresh values evolve is a mystery, but compliance with this current of social evolution is the strongest influence over environmental policy. The result is — or appears to be — an impenetrable tangle of legislation. It was put into memorable prose some 200 years ago by the philosopher William Paley. Here is what he wrote:

> The law in England ... hath grown out of occasion and emergency ... It resembles one of those old mansions, which, instead of being built all at once, after a regular plan ... has been continually receiving additions and repairs suited to the taste, fortune, or conveniency of its successive proprietors. In such a building we look in vain for the elegance and proportion ... which we expect in a modern edifice.

'An impenetrable tangle of legislation', I called it. Yes. But it works, and the contrast in style between this and the elegant patterns of law in some continental countries accounts for a good deal of the controversy we have had over EEC environmental policy. It works, but what we need to examine are the ethical assumptions that lie beneath the laws we make to protect the environment.

REFERENCES

[1] See: Ashby, E. *Reconciling man with the environment.* Oxford, 1978; and Ashby, E. The search for an environmental ethic. *Tanner Lectures on Human Values*, vol 1. 1980, Cambridge.
[2] Report of the Select Committee of the House of Lords on the European Communities, *The Polluter Pays Principle.* Session 1982–83. 10th Report, 1983.

Part 1
THE NATURE AND ASSOCIATED RISKS OF WATER POLLUTION

CHAPTER 1

The origins and pathways of aquatic pollutants

N. J. KING, P. R. HINCHCLIFFE and J. L. VOSSER
Department of the Environment

1.1 INTRODUCTION

The title of this chapter has been changed slightly by substituting 'Pollutants' for 'Pollution' since aquatic pollution is a state that results from the introduction of substances or energy which cause significant and undesirable changes, into the aquatic environment. This chapter discusses, in broad terms, the routes or pathways by which substances which might cause such effects can reach the environment. Although the importance of microbiological pollutants (for example, pathogens from sewage or hospitals) should not be ignored, the discussion here will be restricted mainly to chemical and thermal aspects of pollution.

Discharges may arise from man's activities, or from natural events such as volcanic activity. Origins of pollutants are discussed briefly in the first part of the chapter while the second part deals with the pathways that can be identified.

1.2 ORIGINS

There are many sources of potentially harmful substances which may have significant effects on the aquatic environment. They include:

1. Domestic sewage
2. Industrial effluent arising during the manufacture and use of chemicals
3. Leaching from domestic and toxic waste tips
4. Atmospheric fall out
5. Urban run-off including industrial sites
6. Accidents – spillages and explosions
7. Mineral oil production
8. Mining
9. Farming

10. Solubilisation of solid substances due to changes in the chemistry of the surrounding environment
11. Power generation and other sources of waste heat
12. 'Natural' sources.

Taking these in turn:

1. In terms of quantities discharged, domestic sewage is the major pollutant in the UK. About 95% of the households in the UK are connected to public sewers and a very high proportion (85%) of sewage receives primary and secondary treatment. At least 90% of the polluting load (as biochemical oxygen demand and solids) is removed by efficient treatment, leaving sludge produced by sedimentation and by the aerobic stages of treatment. Thus solving one problem has created another — sludge, which requires treatment and disposal. The importance of proper treatment of sludge and its disposal stems from its propensity to harbour pathogens and to accumulate toxic metals and organic compounds.

The concentration of nitrates in many surface and underground waters shows a continuing tendency to rise. Domestic sewage is an important source of nitrate although increased fertiliser usage is probably responsible for currently increasing concentrations. The process follows a complex pathway taking in plants, their subsequent decay and return to a mineralised state.

2. The constituents of an industrial effluent will reflect the processes used and these effluents may or may not be pretreated before discharge. Requirements for pretreatment and the conditions for discharge of surface waters or to sewer for subsequent co-treatment with domestic sewage at a treatment works will be set by the water authority, having regard to the end-use envisaged for the receiving waters. Pretreatment or full treatment of industrial effluent either alone or with domestic sewage will give rise in many cases to sludge. The composition of both the sludge and the treated liquor will reflect the original effluent and possibly affect the choice of diposal method.

Tables 1.1–1.3 give an indication of the numbers and volumes of discharges (Table 1.1) [1] and the sludge disposal pattern and some analyses for toxic metals and organohalogens in sludge (Tables 1.2 and 1.3 respectively) [2].

3. Domestic and toxic waste tips may well give rise to leachates which can leave the tip and eventually find their way to underground or surface waters. The processes by which leachates are generated include simple washing out of soluble materials by rainfall, decomposition of organic materials to give soluble substances, corrosion and rupture of containers carrying toxic materials. Factors defining the nature, quantity and significance of leachate include the capacity of the materials in the tip to generate water, rainfall and the hydro-geological conditions as well as the composition of materials in the tip.

Proper and effective housekeeping and choice of tipping site will help to reduce the possibilities of leachate escape and subsequent contamination.

Table 1.1
Numbers and volumes of discharges to rivers and canals England and Wales 1975

Discharge	Rivers Non-tidal	Rivers Tidal	Canals	Total	Volume ($\times 10^6 m^3/d$)
Sewage effluent†	4056	348	25	4429	11.2‡
Crude sewage	22	392	0	414	<0.1
Industrial effluent	1692	319	66	2077	7.4
Cooling water	607	135	140	882	59.4
Mine discharges	346	10	10	366	0.9

† About 17% of the total consists of industrial effluent; at least some treatment is given at a sewage treatment works.
‡ Volume excludes that from 25 discharges to canals.

Table 1.2
UK sewage sludge disposal, 1977 data

Sludge production (dry solids) tonnes	1.3×10^6
Disposal pattern	%
Land – Agriculture	44
– Tip	23
Sea	29
Incineration	4

Table 1.3
UK sewage sludge disposal, metals and organohalogens in sludges

Metals (land disposal)	Median concentration mg/kg dry solids
Zinc	1270
Copper	546
Nickel	94
Zinc equivalent	3440
Chromium	335
Cadmium	17
Copper	324
Organohalogens	Range mg/kg dry solids
Dieldrin	<0.05–17.0
gamma-BHC	0.1–0.5
DDT	0.02–0.8
PCB	<0.004–5.0

4. Atmospheric fall out may occur as dry deposition or as a solution in rainfall. Emissions from power plants, industry, traffic etc. will contribute to atmospheric fall out. Under certain atmospheric conditions, photochemical reactions may play an important part in eliminating organic substances or changing their nature and hence toxic properties. The major constituents of atmospheric fall out are, apart from particulate matter, the oxides of sulphur and nitrogen together with minor quantities of other substances. Carbon dioxide is always present due to respiration processes and the combustion of fossil fuels and gives rise to the weakly acidic reaction of rain. Atmospheric fall out may well affect areas a considerable distance from its origin, as 'acid rain' which currently is the subject of international debate. The principal source of acid rain is the combustion of fossil fuels leading to the discharge of oxides of nitrogen and sulphur into the atmosphere, although there remains controversy over the relative contribution of various sources.

Industrial processes can be sources of considerable quantities of metals as particulates in the atmosphere and there may be fall out either in dust or in solution. Over the past few years attempts have been made to quantify the rates of deposition — these results are shown in Table 1.4. They might be regarded as an indicator of the background levels. Much higher levels will of course be found in and around industrial areas.

Table 1.4
Total deposition of metals: arithmetic averages, kg/ha yr

	West Coast		Midland		East Coast	
	Wraymires Cumbria (1971–80)	Plynlimon Powys (1972–76)	Chilton Oxon (1971–80)	Styrrup Notts (1972–76, 1978–79)	Leiston Suffolk (1972–76, 1978)	Collafirth Shetland (1974–76 only)
Pb	0.29	0.22	0.26	0.34	0.14	0.14
Cd	†<0.050	<0.10	<0.30	<0.039	<0.17	<0.025
Hg	†<0.0024	<0.0029	<0.0017	<0.0014	<0.0011	<0.0083
Zn	9.56	0.44	0.50	0.93	1.20	0.44
Cu	0.26	0.29	0.16	0.23	0.12	0.21
Ni	0.12	0.095	0.074	0.061	0.035	0.074
Cr	0.03	0.037	0.019	0.070	0.018	0.017

† *All* values 'less than x'; detection limit varied from site to site and year to year.

Table 1.5 gives some examples of the median concentrations of certain toxic metals in soils and dusts in urban and rural locations. The higher values in urban soils reflect amongst other things industrial activity.

Table 1.5
Concentration of four toxic metals in soils and dusts

Source	Concentration medians (ppm)			
	Pb	Cd	Cu	Zn
Soil (Urban)†	436	1.8	74	397
(Rural)‡	<40	<1	<30	100–200
House dust (Urban)†	637	7.5	203	1129
(Rural)§	498	8.2	—	—

† Based on preliminary results for 10 towns in the Imperial College 53-town national survey.
‡ Based on stream sediment survey of England and Wales by Imperial College (Wolfson Geochemical Atlas).
§ Limited data based on N. Petherton only. No other rural dusts yet reported for the UK.

5. Urban run-off will consist of water draining from streets, buildings, and industrial premises. Its composition will reflect its origins. For example run-off from roads would be expected to contain traces of hydrocarbons, including polynuclear aromatics (PAH), lead compounds, cadmium from tyres, particles of rubber and anything else that is present which the water could dissolve or entrain and suspend.

6. Accidents such as spillages and explosions may give rise to pollution but, except in rare cases of catastrophic proportions, they are generally local in effect.

7. Mineral oil production and storage may be sources of hydrocarbons irrespective of whether the oil is produced on or off-shore. It is important to remember that pollution by mineral oil does not only arise from the blanketing effect of oil on water but also from the small but significant solubility of toxic hydrocarbons in water and the effect of the heaviest fractions which sink and settle on the sea or river beds. Exploration for oil also involves the use of special muds or fluids to facilitate the drilling operation; these aids may contain additives which may be toxic.

8. Mining operations inevitably involve the production of spoil which may be removed from the mine and dumped. Leaching and washing processes will give rise to suspended matter and, depending on the composition of the spoil, solutions of potentially harmful materials. Pollution may arise from the discharges of coal mine water, coal washery and tailing plants, containing large amounts of suspended matter, to streams and water courses. Operating and disused mines may be a

source of acidic water containing iron and other salts depending on the composition of the coal-bearing strata.

As far as other types of mining operations are concerned, waters from these would usually contain traces or more of whatever was being mined. With clay or china clay workings the problem will almost be entirely that of very large quantities of suspended matter rejected by washing and particle size classification process.

9. Agricultural chemicals are deliberately released into the environment to cause change. There is wide variation, both in methods of application of the chemicals, and in types of site where they are likely to be used. In the case of, for example, a common fertiliser, dispersion may be over a very wide area. On the other hand, a specialised pesticide may be released into a very localised area. From the application site, typical routes into the environment include run-off into rivers or groundwater, or direct uptake by organisms at the site of application.

Although Codes of Practice exist for dealing with farm waste disposal, farming activities may also give rise to pollution of water from silage liquor, crop spraying, animal wastes, or the careless disposal of not quite empty containers of pesticides, herbicides or disinfectants.

10. Solubilisation of solid or insoluble substances by changes in the chemistry of the surrounding environment can be exemplified by the mobilisation of metal salts, contained in sludge disposed on land, by changes in pH value of the soil. Anaerobic processes within the depth of tipped wastes can generate acids which then attack metallic wastes and give rise to leachates which in turn may find their way to surface or underground waters. Acid rain falling on soil can also act to solubilise metals, leading to pollution of surface and possibly underground waters.

11. Apart from the atmospheric pollution discussed above, power generation can also cause problems by the discharge of hot water to natural surface waters. Thermal pollution may upset the balance of local ecology in a number of ways. For example, raised temperatures may directly stress some organisms, or may result in the deoxygenation of water by increasing the rate of degradation of organic matter.

12. When considering the fate or effects of a pollutant, or even when trying to assess the health or quality of an ecosystem, it is necessary to take into account natural modifying factors which may exist in the environment. The modifying factors may include substances which affect biota — for example, humic acids from peat bogs, or heavy metals from ore deposits, and it might be convenient to consider them as 'natural pollutants'. In reality, they are merely natural constraints on the ecosystem, but nevertheless, by stressing the system, they may have significant influences on the way in which the ecosystem responds to

anthropogenic pollutants. 'Natural' pollution might result from the following situations:

1. Changes in the use of land, e.g. afforestation and deforestation, the former leading to acidic conditions from coniferous trees, the latter increasing the tendency of the land to flood;
2. Volcanic action and forest/brush fires — these are sources of acidic materials and other potentially noxious substances;
3. Lightning which can be a source of nitrogen oxides which can reach surface waters by rainfall;
4. Eutrophication — this is the progressive enrichment of natural water courses principally by nitrogen and phosphorus followed by algal blooms, their death, and the subsequent deoxygenation of the water. Although the main sources of the nutrients tend to be anthropogenic, natural sources, such as large colonies of birds near water courses or reservoirs, may be a contributory factor.

1.3 PATHWAYS

When considering the path of a pollutant through the environment, it is necessary to take into account its source, discharge route, likely receiving environments, and its physical, chemical and biological behaviour at all stages of the journey. It will not always be necessary to identify a final resting place for the chemical — the chemical may degrade to negligible concentrations *en route*, or it may be obvious that it will cause problems along its pathway, unless controlled to a level which will prevent significant quantities reaching the final sink.

Although at first glance there may seem to be myriad possible pathways from source through the environment, in practice, by considering worst likely cases, and taking into account practical limitations on mobility, the number of possibilities becomes fairly manageable. The routes we will most often be concerned with will generally start off with aqueous discharge to surface water or sewer, or spreading or tipping to land, or discharge to the atmosphere. We must then consider transport in aqueous solution or suspension through surface or groundwaters, absorption onto sludge or sediment, leaching from tips or soil, and aerial transport and deposition by rain. Migration from one medium to another may be important, particularly at the water—sediment interface. Uptake by biota and incorporation into the food chain must also be considered, particularly in the case of a persistent or accumulative material.

So it can be seen that, although the pathway of a given pollutant from its source to the aqueous environment may be simple, for example, direct discharge, or very tortuous, in general there are two processes which will have a significant effect on the length and complexity of the path. They are, (a) transfer, and (b) transformation, and one or both may operate during the movement of the pollutant through the environment.

Transfer may be illustrated by emissions to the atmosphere of sulphur dioxide in, say, a flue gas followed by its dissolution in atmospheric moisture and subsequent deposition on land or water in rainfall — it has been transferred from one environment sector to another. An example of transformation is provided by the dumping of an industrial waste containing toxic metals hydroxides which, in the dumping site, come into contact with an acidic waste. This would bring the metals into solution and there is a possibility that leaching out of the tip may occur with the liquor finding its way into an underground water. Thus what might have been a short pathway to a sink has been lengthened by transformation possibly leading to a hazard elsewhere.

Other processes such as precipitation, leaching and volatilisation can transfer pollutants into and out of water. The physical process of absorption plays an important part in the transfer of pollutants, many substances are strongly absorbed onto sewage sludge and other solids and may pass to land instead of being discharged to surface waters in sewage works or other effluents.

The transformation process may include any one or more of the following reactions:

Photodegradation
Biodegradation (aerobic and anaerobic)
Hydrolysis
Neutralisation
Precipitation
Solubilisation
Chelation

which may occur during the passage of a pollutant from its source to the aquatic environment either incidentally or deliberately in treatment plants to which the pollutant is discharged. In the case of the degradation reactions the effect is to break down the molecule to simpler structures and in some cases leading to complete mineralisation. The result is that the concentration of the pollutant is reduced to leave a small amount of residual material.

Of the other reactions, solubilisation may be most significant in the case of spoil heaps and tips. Substances which have very low solubilities may be slowly dissolved or leached out by rainfall, by liquor produced from tipped material or by microbiological activity. Then, by drainage through faults in the sub-surface strata, the pollutants find their way to natural waters.

Having briefly described the processes which may affect the pathways, it may be useful now to describe the principal pathways that may be found and how they may be changed by the processes described above *en route* from the source to the point of discharge into the aquatic environment.

1. Direct discharges to inland surface water of untreated effluents either with or without the presence of untreated domestic sewage are rare in the UK. The pathway provided is of course the simplest and unlikely to be affected much by the

transfer and transformation processes until the discharge reaches its destination. The rarity of this type of discharge is the result of the enactment of provisions designed to prevent pollution of rivers and surface waters (for example, 1951 and 1961 Acts and the Control of Pollution Act 1974 Part II).

Direct discharges are more common to estuarial and coastal waters either by short or long outfalls where the dilution of the discharge by the receiving water is likely to be high with good mixing and dispersion by currents and tides. Unfortunately, in several instances the design has been less than perfect and conditions in the receiving water have changed with the passage of time, resulting in situations where the contents of the discharge are not properly carried out to sea and return to the shore either causing problems of an aesthetic nature or contaminating fisheries.

The implementation of the appropriate sections of the Control of Pollution Act should help to reduce these incidences of coastal pollution.

2. In many instances partial treatment is given to the waste waters before discharge — this usually consists of screening to remove large objects from the flow, followed by sedimentation to remove smaller suspended particles. With proper design it should be possible to remove about 65% of the suspended matter as sludge and at the same time some 30% of the biochemical oxygen demand. By interposing the partial treatment stage, the pathway has been divided because a sludge has been produced and which needs disposal. Thus there are two routes to surface waters, direct discharge of the supernatant liquor, and the less direct route via sludge. Either route could lead to the possibility that adverse effects may arise from inappropriate disposal.

3. The use of full treatment of wastewater will produce a division of the pathway in the same way as partial treatment. Full treatment is normally given when the discharge from the treatment plant is made to inland, non-tidal waters such as rivers. If the sludge produced is taken to sea for disposal then obviously the pathway to pollution inland no longer exists and is replaced by one which may involve the leachate from the dumped sludge being dispersed into the sea.

The possibility of establishing a pathway to natural waters as a result of the land disposal of liquid sludge has been mentioned, but there are many sewage treatment works which dewater their sludges before disposal. In these cases depending, on the dewatering process used, the moisture content can be in the range of 60 to 80%. At the lower end of the range the dried sludge may contain inorganic salts used as conditioning agents prior to dewatering. Before it can be disposed of it may have to be broken up prior to ploughing into land. Sludge cake of this type should reduce the possibility of pathways being created after dumping or ploughing into land because it does not take up water to any great extent, thus considerably reducing the possibility of leaching. Only major changes in chemistry would be likely to promote leaching. Sludges of higher moisture content possibly increase the chance of a pathway being more readily established.

Ch. 1]   **The Origins and Pathways of Aquatic Pollutants**   31

A pathway for untreated sewage and wastewaters which may have significance for subsequent pollution of underground and surface waters is that opened up by leaking sewers, cesspits and septic tanks. Wrong connections between surface water and foul sewers may also lead to polluted water. Human error should never be discounted!

4. Spoil and other tips, for example, for household and toxic waste, may well lead to problems by the leaching of soluble materials by rainfall. Depending on the nature of the site, leachate may percolate directly to groundwater or run out of the tip into streams or ditches. If the tip is situated over heavily fissured strata then it is possible for leachate to travel a considerable distance underground and possibly reach a water source which might have been thought safe from contamination.

5. Emission to the atmosphere from power generation and industrial processing plants may contain materials which may fall to the ground as 'dusts' which are then dissolved either by the water into which they have fallen or by rainfall. Alternatively some of the constituents of the emission may fall to the ground already dissolved in rainfall and enter surface or underground waters in this way. In particular, the initial pathway for acid rain is atmospheric dispersion and transport followed by deposition in rain and dust. Effects on vegetation receiving the precipitation may be observed at this stage, but some of the acid rain will be carried further into streams and lakes, possibly causing problems for aquatic life.

Volcanic eruptions may have consequences for surface waters both near and far from the volcano; in general the pathways will be similar to those for manmade atmospheric emissions.

1.4 CONCLUSION

This chapter has attempted to show why water is an important pathway for pollutants, and why the aquatic sector is the site of important effects. Most pollutants will find their way into surface water at some stage after their release into the environment, and for many it will form the initial pathway. For this reason there is considerable emphasis in environmental protection on investigating and controlling the origins and pathways of aquatic pollutants.

1.5 ACKNOWLEDGEMENT

This chapter is reproduced by permission of the Controller of Her Majesty's Stationery Office.

© Crown copyright 1983.

1.6 REFERENCES

[1] River Pollution Survey of England and Wales. Updated 1975. *River Quality and Discharges of Sewage Effluents.* HMSO, London, 1978.
[2] Report of the Sub-Committee on the Disposal of Sewage Sludge to Land. Standing Committee on the Disposal of Sewage Sludge. Report No. 20. Department of the Environment, June, 1981.
[3] UKAEA Harwell Reports: Survey of Atmospheric Tracer Elements in the UK. 1972–1973 AERE Report R7669; 1974 AERE Report R8083; 1975 AERE Report R8393; 1976 AERE Report R8869; 1977 AERE Report R9164; HMSO, London.

CHAPTER 2

The nature and concentration of pollutants in water

M. FIELDING and D. T. E. HUNT
Water Research Centre

2.1 INTRODUCTION

Before one can discuss the nature and concentration of pollutants in water, some definition of pollutant is needed. Various approaches are possible; on the one hand, we have definitions – usually implied rather than explicit – which suggest that any anthropogenic substance detected – regardless of adverse effects – is a pollutant, while on the other hand we have definitions which link pollutants exclusively to both anthropogenic origins and adverse environmental effects. The latter type are more generally useful and, while accepting that no approach is unassailable, we have used the following definition of water pollution (taken from the EC Directive on pollution by 'dangerous substances discharged into the aquatic environment' [1]) as a guide:

> ... the discharge by man, directly or indirectly, of substances or energy into the aquatic environment, the results of which are such as to cause hazards to human health, harm to living resources and to aquatic ecosystems, damage to amenities or interference with other legitimate uses of water.

Some important points require to be noted, however:

(i) The available data, on both organic and inorganic substances in water, are not neatly categorised into 'natural' or 'anthropogenic', 'harmful' or 'harmless'.
(ii) Whether or not a substance causes deleterious effects will usually depend upon its concentration. Additionally, and very importantly, there is much contention over whether or not some substances have adverse effects at the concentrations observed, and there are many substances for which any adverse effects are potential and ill-defined.

Thus, 'pollutant' inevitably must include 'potential pollutant' in this chapter.

(iii) In order to understand pollution it is difficult to ignore natural constituents and their sources.

Certain substances, not introduced by man, may cause undesirable effects, for example natural substances that produce tastes and odours in water supply.

The chapter restricts itself to chemical pollution and therefore pollution by energy, radioactivity and living organisms is not included.

Thus, the task has been to cover the nature and concentration of pollutants in both fresh and marine waters. Data on these and associated aspects are abundant and in a chapter of this size only a general picture of the current situation has been feasible.

Some form of classification of chemical pollutants is needed and therefore we have chosen to classify, where appropriate, according to the chemical nature of the substances with further sub-division by water type.

2.2 INORGANIC POLLUTANTS

2.2.1 Acids and alkalis

The metabolic processes of aquatic organisms function optimally over particular pH ranges and species diversity will be markedly reduced as the pH tends to extreme values.

The influence of acids or alkalis is not, however, restricted to direct biological effects. Changes in pH may profoundly affect chemical processes and thus cause indirect pollution by liberating — or rendering more harmful — substances already present. (As a corollary, the behaviour and effects of many pollutants may be greatly influenced by the pH and composition of the receiving water.)

Gross pollution by acids or alkalis can arise from localised industrial sources — for example, acid mine drainage [2,3]. Of greater topical interest — and wider potential impact — is acidification of poorly buffered waters, often attributed to 'acid rain' from oxides of nitrogen and sulphur discharged to the atmosphere by industrial processes.

Reported effects include decline or extinction of fish stocks, reduction in biological diversity and increased aluminium concentrations [4,5]. Although acidification may itself damage fish (reference [6], pp. 21–45), indirect effects — notably raised aluminium levels — may also be implicated in the decline of stocks [4].

Although acidification is often attributed to acid rain, the 'acid budget' of an area involves many different processes. Thus land usage may be important, and there is evidence [7] that the pH of streams in conifer forests tends to be lower than that of similar streams in adjacent, unafforested areas. It should also be noted that pH determination in poorly buffered waters of low ionic strength is difficult [8], and examination of pH trends may be prejudiced by the large errors which can arise.

2.2.2 Nitrogen species and phosphate

Both nitrate and phosphate are potential causes of eutrophication — excessive algal growth, which may lead to deoxygenation and sulphide formation. Phosphate is normally more important, and its concentration in rivers has risen with increased use of biodegradable detergents. The onset of eutrophication depends on many factors, but it often accompanies phosphorus concentrations above about $35 \mu g$ P/l in lakes and reservoirs [9].

Nitrate concentrations in rivers have also risen, sewage effluent and land run-off being likely causes. This is of concern because nitrate in drinking water has been associated with methaemoglobinaemia ('blue baby') — a rare disorder of bottle-fed infants. The EC Directive on the Quality of Water for Human Consumption sets a Maximum Admissible Concentration (MAC) for nitrate of 50 mg NO_3/l.

Free ammonia, NH_3, is toxic to fish at relatively low concentrations (*ca.* 0.2 mg NH_3/l), but it is in equilibrium with the essentially innocuous NH_4^+ ion and their relative proportions (and, therefore, the toxicity of a given *total* ammonia concentration) will depend upon pH (reference [6], pp. 85–102). Ammonia also interferes with water treatment disinfection.

2.2.3 Trace metals, metalloids and other trace elements

A concentration of 1 ppm (1 mg/l) is usually regarded [10,11] as the demarcation between 'major' and 'trace' (or 'minor') constituents. However, the concentrations of trace elements in waters are usually highly variable (in space and time) and may exceed 1 mg/l in certain instances. Examples of elements in the three categories are:

Trace metals:	Al, Fe, Mn, Pb, Cu, Cd, Ni, Zn, Hg, Cr, Co, V, Sr, Mo, Ag, Sn, Tl;
Trace metalloids:	As, Se, Sb;
Other trace elements:	B, Br, I, N, P, F.

The micronutrients (N and P) have already been considered, and Br, I and F are not widespread water pollutants. Attention will, therefore, be restricted to trace metals and metalloids. Interest in these has grown markedly in the last 20 years, because of the deleterious effects which they are known, or suspected, to cause. Many of the elements are, however, biologically essential and become toxic only in excess. In general, the toxicity of the 'non-essentials' (Batley [12] cites Pb, Cd, Hg, Ag, Tl, As) is greater.

Water quality standards have been set for many trace elements, for the protection of both humans and other organisms. Standards are dealt with in Chapter 7 of this book [13], and we will merely note that they often specify very low concentrations (*ca.* 1 to 100µg/l) for trace elements, particularly for the protection of aquatic life.

Although the following section deals with trace elements in water, the behaviour of such elements in soil is also of interest to the water industry, because of their enrichment on soils amended with sewage sludge.

The sources and pathways of pollutants in the aquatic environment are discussed in Chapter 1 of this book [14]. It will simply be noted that those for trace elements are legion, reflecting the widespread use of these substances and their complex environmental behaviour.

2.2.3.1 *Forms of trace elements in water*

Trace metals and metalloids can exist in water in a great variety of forms (see Fig. 2.1), and it has been increasingly recognised [10, 15–17] that a knowledge of the distribution of the elements between these forms ('speciation') will often be required for an understanding of their behaviour and effects. A knowledge of trace element speciation may be of benefit in the following areas:

(i) Toxicity to aquatic organisms. This may be heavily dependent upon the form of the element (references [18], [19] and [6], pp. 189–220).

(ii) Water and wastewater treatment. The solubility of trace elements (and therefore the efficiency of their removal by precipitation) depends not only upon the inherent solubility of the solid, but also upon the solution speciation [20, 21].

(iii) Acquisition of trace metals in distribution systems. This is often dependent on the solubility of pipe deposits, and hence (as in (ii)) on the solution speciation. The WRC has used thermodynamic calculations of lead solubility to aid research into water treatments to reduce plumbosolvency [20, 22].

(iv) Pollution control. The transport of trace elements in water bodies may depend markedly on the speciation – in particular, on the relative magnitudes of the particulate and dissolved fractions [10].

The characterisation of trace element speciation poses enormous difficulties. Though useful, thermodynamic calculation has some serious limitations [20] and analytical approaches are normally capable only of dividing the total concentration into operationally defined fractions [10, 12]. However, Fig. 2.2 shows how a knowledge of speciation (albeit imperfect) can be beneficial – in this example, by avoiding serious misconceptions about the solubility of the element and its variation with changes in water quality.

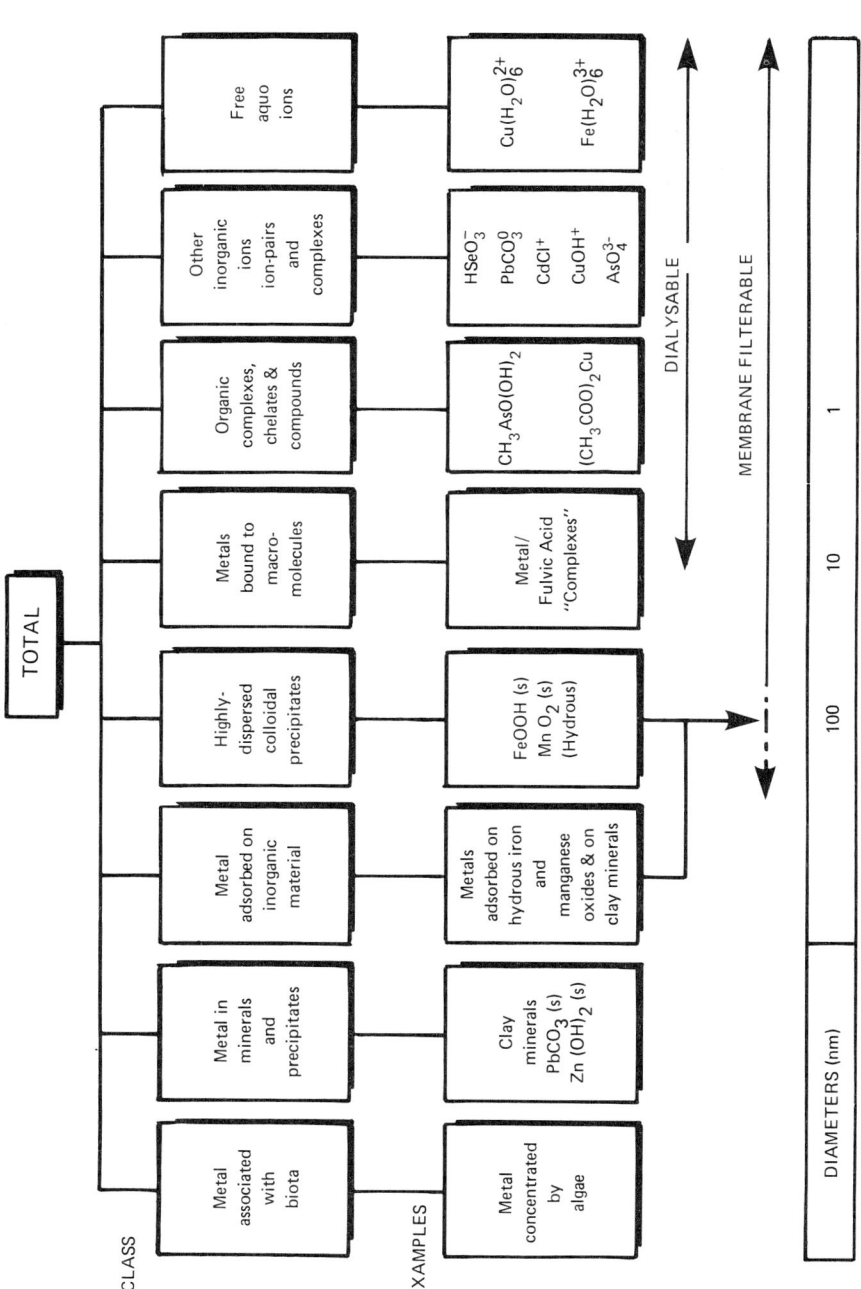

Fig. 2.1 — Classification scheme for trace metal (metalloid) species in aqueous systems (based on a similar scheme in reference [15]).

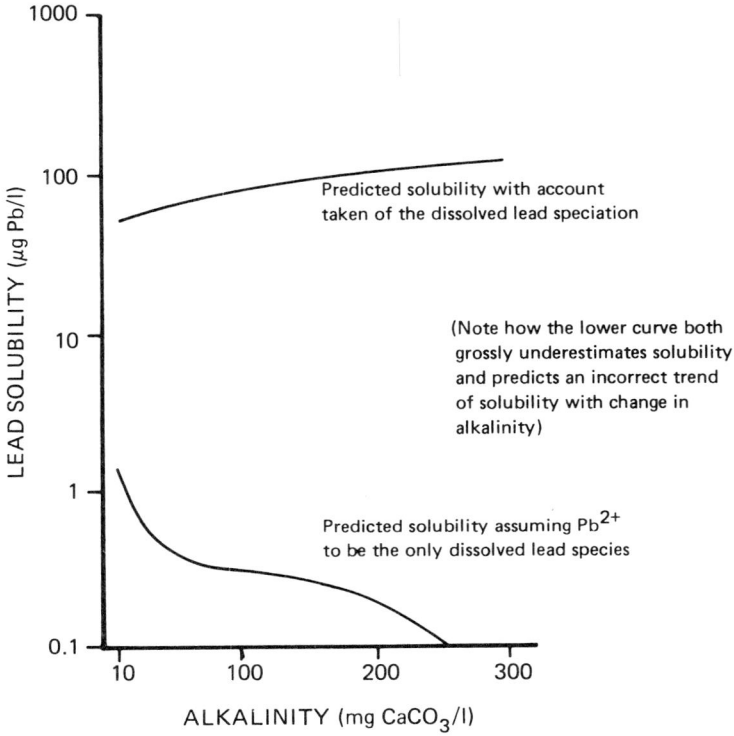

Fig. 2.2 — Predictions of lead solubility as a function of alkalinity at pH 8.5, (adapted from reference [20]).

2.2.3.2 *Concentrations of trace metals and metalloids in water*

Tables 2.1 and 2.2 provide a very limited summary of data on concentrations. It is emphasised that the extreme variability of the concentrations of many trace elements (in space and time) precludes the use of such data except as a general guide. Certain points are noteworthy, however:

  (i) With respect to drinking water, the influences of water treatment (Al in soft waters) and of the plumbing system (Cu, Zn, Pb) can be readily discerned.
 (ii) Dissolved trace metal concentrations in the open sea are very low ($<1\mu g/l$ in most cases). Coastal values are much higher, particularly near major estuaries, as a result of both natural and anthropogenic inputs.

Ch. 2] The Nature and Concentration of Pollutants in Water

Table 2.1

Trace metal concentrations in river waters and drinking waters (all data in µg/l)

	Metal concentrations in rivers[†] (Data from rivers throughout the world)		Mean metal concentrations in drinking waters[‡] (Data from UK drinking waters only)			
			Soft water towns[¶]		Hard water towns[∥]	
Element	Range	Central value[§]	First-draw	Fully-flushed	First-draw	Fully-flushed
Aluminium	0.1–10 (D)	1 (D)	190 (T)	180 (T)	26 (T)[††]	27 (T)[††]
Cadmium	0.2–20 (D)	3 (D)	2 (T)[††]	2 (T)[††]	2 (T)[††]	2 (T)[††]
Chromium	0.1–10 (D)	0.5 (D)	3 (T)[††]	3 (T)[††]	5 (T)	4 (T)
Cobalt	0.1–10 (D)	0.5 (D)	5 (T)[††]	5 (T)[††]	5 (T)[††]	5 (T)[††]
Copper	0.5–50 (D)	7 (D)	140 (T)	24 (T)	300 (T)	14 (T)
Iron	3–300 (D)	40 (D)	110 (T)	100 (T)	40 (T)	31 (T)
Lead	1–100 (D)	5 (D)	47 (T)	11 (T)	14 (T)	3 (T)
Manganese	1–100 (D)	5 (D)	20 (T)	22 (T)	4 (T)	5 (T)
Mercury	<1 (T)	<0.5 (T)	—	—	—	—
Molybdenum	0.2–20 (D)	5 (D)	37 (T)[††]	37 (T)[††]	45 (T)[††]	54 (T)[††]
Nickel	0.5–50 (D)	10 (D)	20 (T)[††]	18 (T)[††]	18 (T)[††]	16 (T)[††]
Silver	0.01–1 (?)	0.1 (?)	3 (T)[††]	3 (T)[††]	3 (T)[††]	3 (T)[††]
Vanadium	0.1–10 (D)	3 (D)	4 (T)	4 (T)	12 (T)	12 (T)
Zinc	2–200 (D)	20 (D)	63 (T)	19 (T)	110 (T)	9 (T)

[†] From reference [10].
[‡] From reference [23]. Where individual results were reported below the limit of detection, they were assigned a value of half the limit of detection. In cases marked [††], the overall mean value shown was also below the limit of detection.
[§] Subjective judgement, not statistical (see reference [10]).
[¶] Hardness <120 mg CaCO$_3$/l.
[∥] Hardness >200 mg CaCO$_3$/l.
(D) – Dissolved; (T) – Total; (?) – Fraction not specified.

Table 2.2

Trace metal concentrations in UK coastal waters and the North Sea
(all data in $\mu g/l$)

Element	Range of concentrations (lower end refers to the northern North Sea)[†]	
Cadmium	0.2–0.025	(D)
Copper	2–0.2	(D)
Mercury	0.02–<0.003	(R)
Nickel	4–0.25	(D)
Zinc	10–1	(D)

[†] Data summarised from maps of trace metal distributions presented in reference [24].
(D) – Dissolved; (R) – Reactive.

(iii) Estuarine concentrations of dissolved trace metals have been excluded, because of a lack of collated data. Typically, they will lie between those quoted for river water and coastal sea water, though they may be higher than expected from the relevant fresh water and salt water concentrations (by virtue of estuarine inputs) or lower (by virtue of removal from solution during estuarine mixing).

(iv) Most of the data presented for rivers and marine waters relate to the dissolved fraction. Concentrations in the 'suspended particulate' fraction may be higher or lower, and will depend markedly upon the total suspended matter concentrations, and thus on the prevailing sedimentary and hydrological conditions. Different elements may be associated to different extents with the dissolved and suspended particulate phases [10].

2.2.4 Sulphur

Sulphur is not normally of concern as a direct pollutant, but – in the form of sulphide – is often of concern as a product of anoxic conditions. Such conditions usually arise as a result of high loadings of oxygen-consuming organic matter, which may in turn be of natural or anthropogenic origin. Anthropogenic anoxia may be direct (for example, from sewage effluent) or indirect (for example, from eutrophication). Although sulphide is itself toxic, it can also give rise to indirect toxic effects by mobilisation of trace metals bound, under oxygenated conditions, to particulate matter.

2.2.5 Miscellaneous

A variety of inorganic pollutants, other than those considered so far, may also be introduced to the aquatic environment by man's activities. Examples are cyanide (from metal plating industries) and sulphite (from paper manufacture). Commonly, any effects of such pollutants are highly localised.

2.3 ORGANIC POLLUTANTS

Natural waters contain a complex mixture of organic substances covering a diverse variety of chemical class. Sources of natural organic substances may be aquatic (that is, carbon fixed within the aquatic environment) or non-aquatic (that is, leached from land vegetation and soils). In addition, no less complex mixtures of anthropogenic substances are introduced by industrial and domestic effluents, surface drainage, spillage and direct addition (for example, aquatic herbicides). Less obvious sources exist, such as synthesis during disinfection, leaching from distribution mains, atmospheric deposition and rainfall.

Distinction between natural and anthropogenic substances is often difficult because some 'industrial' chemicals have natural origins, for example, production of halogenated hydrocarbons, such as chloroform and carbon tetrachloride by red algae [25], some anthropogenic substances are natural materials, while others are impurities or by-products not immediately recognisable as industrially used chemicals.

Organic substances in water are in a constant state of biotic and abiotic modification. Thus a harmful chemical may be altered to an innocuous product (possibly vice versa). Substances which are refractory may become dramatically accumulated at certain stages of the food chain — and become a serious pollutant.

Organic pollutants may be associated with a variety of environmental problems, usually of a biological nature, such as toxic effects on aquatic life and risks to public health. However, odours and tastes imparted by organic chemicals, sometimes of natural origin, can be highly troublesome.

A vast number of organic chemicals are known but only around 10 000 or so are manufactured in relatively large numbers, say greater than 500 kg per annum. Compared to this, the number of widely accepted water pollutants is very small indeed. In terms of available data on organic substances in water we have, on the one hand, information on a small number of largely undisputed water pollutants and, on the other hand, an enormous quantity of data on organic substances in general. For example, the COST 64B inventory [26], which excludes marine waters, has 20 000 entries on over 3000 organic substances. This mass of information has been summarised by reviewing natural and anthropogenic organic substances in general and several widely accepted chemical pollutants.

2.3.1 Total organic matter

The total organic matter in water is made up of dissolved organic substances and organic substances associated with suspended particulate matter. When estimated, the particulate matter is arbitrarily defined by a filtration stage in the sample processing, for example a $0.45 \mu m$ filter. Clearly the total organic matter in water will reflect variations in the amount of particulate matter in suspension, and therefore occasionally, for example in rivers in high flow conditions, the total can rise dramatically. Table 2.3 summarises total amounts of organic matter in unpolluted to slightly polluted water under average conditions.

Table 2.3
Total organic matter in water (mg as carbon/l)

	Particulate organic carbon	Total organic carbon (dissolved plus particulate)
Marine water	0.006–0.215 (a)	0.6–1.84 (a)
Freshwater	10% of TOC (b)† 28% of TOC (c)†	1–20 (d)
Drinking water	negligible	0.5–20 (d)

† River water.
(a) = reference [27]; (b) = reference [28]; (c) = reference [29]; (d) = unpublished data (WRC).

Thus, total amounts are relatively low and range from barely detectable in some groundwaters (and deep seawater), to much higher levels in peaty, upland surface waters. Higher levels still (>100 mg as carbon/l) are possible in some situations, for example, very productive or swamp waters and badly polluted water.

Although most information on composition of the organic matter relates to dissolved components, some exists for marine [30] and lake [28] particulate matter.

2.3.2 Natural organic substances

Most studies on the identity of natural organic substances relate to marine or lake water and good reviews exist [31, 32]. Hunter and Liss [33] give typical levels for marine waters (Table 2.4).

Table 2.4
Average levels of natural organic substances in marine water

	µg as carbon/l		µg as carbon/l
Free amino acids	10	Vitamins	0.006
Combined amino acids	50	Ketones	10
Free sugars	20	Aldehydes	5
Combined sugars	200	Hydrocarbons	5
Fatty acids	10	Urea	10
Phenols	2	Uronic acids	18
Sterols	0.2	Uncharacterised	660

Ch. 2]   **The Nature and Concentration of Pollutants in Water**   43

A review by Cranwell [28] gives data for lake water for several of the groups in Table 2.4; levels being higher, especially for productive lakes, but usually within an order of magnitude.

Each of the groups contains many substances; some, such as the hydrocarbons, can be highly complex. Major constituents of fresh waters are humic substances, which in highly coloured upland waters may amount to 30 mg/l or more.

The content of natural organic substances in river water is not well established and a major complication is introduction of such compounds via sewage effluent. Waggott [34] has given data for a typical domestic sewage effluent (Table 2.5).

Table 2.5
Average levels of natural organic substances in domestic sewage effluent

	µg/l		µg/l
Free amino acids		Polysaccharides	
(as leucine)	1440	(as glucose)	1700
Proteins (as leucine)	490	Steroids (as cholesterol)	165
Carbohydrates		Acids	2810
(as glucose)	100	Humic substances	17200

A review of natural substances in drinking water is not available but drinking water derived from a source containing a high proportion of sewage effluent could have levels of about an order of magnitude lower than those in Table 2.5.

The above summary hides to some extent the surprising diversity and complexity of natural organic substances which is emerging. It is difficult to ignore such substances for the reasons given already, and because of the possibility that some can 'trap' or solubilise pollutants, for example, humanic substances, or react during disinfection and synthesise chlorinated organic pollutants (see below).

2.3.3 Anthropogenic substances
2.3.3.1 *Organic substances in general*
Several papers refer to the fact that prior to about 1970 less than 100 organic compounds had been identified in water and only a handful in drinking water. Today over 2000 have been characterised. The sensitive analytical techniques, such as gas chromatography–mass spectrometry, which have made this possible relentlessly reveal the presence of a wide range of anthropogenic substances in almost all types of natural water. The following examples illustrate the existing information for certain types of water but detailed information on organic pollutants in water exists [35–40].

UK drinking waters examined by James et al. [37] using GC–MS techniques, revealed a complex range of chemicals regardless of raw water source. In 14 samples, a total of 323 different chemicals were identified. The majority were probably of anthropogenic origin but a clear-cut distinction was difficult. Table 2.6 summarises the findings.

Table 2.6
Compound types identified in UK drinking water

Hydrocarbons	Carboxylic acids and esters
saturated/unsaturated aliphatic (68)	acids (15)
alicyclic (5)	esters (31)
aromatic (48)	
polynuclear aromatic (20)	
Halogenated compounds	Oxygenated compounds
aliphatic (53)	ketones (19)
aromatic (8)	aldehydes (7)
ethers (4)	ethers (11)
miscellaneous (14)	miscellaneous (17)
Miscellaneous compounds (23)	

Numbers in parentheses are total numbers identified.

The specific conclusions from the findings, which are typical of such surveys, were as follows:

- all samples revealed a very complex mixture of organic substances although drinking water derived from river water containing sewage effluent generally contained higher concentrations of many compounds;
- the concentration of most chemicals was $<1\mu g/l$ and only chloroform approached $100\,\mu g/l$;
- several known or suspected organic chemical carcinogens were found in drinking water;
- most of the compounds have never been evaluated in terms of safety;
- the data only relates to a small proportion of the total organic substances present in drinking water. Consequently, conclusions on overall water quality need to be tempered accordingly.

The overall conclusion was that much of the population is exposed, via drinking water, to minute quantities of a wide range of organic chemicals, some of which, at much higher concentrations, would give grave concern.

When similar analytical techniques are applied to surface waters, particularly river waters, similar complexity and diversity is usually encountered. Table 2.7 summarises a survey of a UK lowland river basin (including sewage effluents) used for potable water supply. In this survey [38] the compounds identified were arranged according to their possible source or, when this was not feasible, chemical class. The concentrations of most substances were estimated to be $<0.1\mu g/l$.

Table 2.7
Compound types identified in UK lowland river catchment area

Antioxidants (1)	Surfactants, additives and metabolites (30)
Corrosion inhibitors (several)	
Flame retardants (1)	Organic acids (3+)
Flavour and fragrance compounds (6)	Organic bases (5+)
Germicides and preservatives (5+)	Chlorinated compounds (14)
Oils (various fuel and lube oils)	Ethers (2)
Pesticides and herbicides (11+)	Organic sulphur compounds (5)
Plasticisers (10+)	Polynuclear aromatic hydrocarbons and aryl alkanes (16+)
Pharmaceuticals and metabolites (6)	
Solvents (24+)	

Numbers in parentheses are numbers of compounds identified.
+ means unspecified numbers of isomers, for example phenyloctane isomers.

In more polluted surface waters many additional substances may be encountered and overall concentrations may be considerably higher.

Complex mixtures of anthropogenic organic chemicals can be revealed in marine water, especially coastal and estuarine water directly or indirectly receiving wastewater. For example, Gschwend *et al.* [41] studied likely inputs and seasonal variations of a complex mixture of organic chemicals (usually present at $<0.1\mu g/l$) including alkylaromatics, alkanes, aldehydes, organic sulphides and many miscellaneous substances. Anthropogenic inputs were found to be a major factor for many of the substances.

The main difficulty and general lack of understanding with anthropogenic organic chemicals in natural waters is less to do with acute effects on aquatic and human life, as with potential long-term effects of low concentrations — which may be unspecific and difficult to detect. In practice, the relatively recently acquired analytical ability to demonstrate the presence of low concentrations of a diverse variety of anthropogenic organic chemicals in natural waters is countered to some extent by an inability to evaluate satisfactorily their practical significance.

2.3.3.2 *Specific organic pollutants*

Some organic chemicals can find their way into water and produce distinct adverse effects under certain circumstances and therefore are considered by many to be unquestionable pollutants. Some examples follow.

Hydrocarbons

Pollution by hydrocarbons results from spillage of crude oil and crude oil products, and general industrial and domestic usage of oil products, such as fuel oils and lubricating oils. Additional inputs arise from incomplete combustion of organic materials — mainly fossil fuels. Anthropogenic hydrocarbons encountered in water are usually complex and include aliphatic-, alkylaromatic-, aromatic- and polynuclear aromatic hydrocarbons. Except for the lower aromatic hydrocarbons, concentrations in water tend to be limited by poor solubility and most of the pollutional load becomes associated with particulate matter, sediments or other soluble organic matter. The literature — mainly on marine hydrocarbon pollution and its effects — is extensive.

Table 2.8 summarises data on hydrocarbons in marine environments reported by Farrington and Meyers [42].

Table 2.8
Levels of hydrocarbons in marine water, organisms and sediments

Marine water	$1-100 \mu g/l$
'Tar particles'	$1-40 \mu g/l$
Organisms	$1-200 \mu g/g$ wet weight
Organisms (oil polluted)	up to $545 \mu g/g$ wet weight
Sediments (continental shelf)	$1-100 \mu g/l$ dry weight
Sediments (estuarine)	$5-100+ \mu g/g$ dry weight
Sediments (oil polluted, coastal)	up to $12400 \mu g/g$ dry weight

Hydrocarbon pollution of freshwater by spillage of refined oils is a common occurrence which causes many problems, usually of a short-term nature. Background levels of contamination in surface waters, especially rivers, are not firmly established. One potentially serious problem is the contamination of the ground due to accidental spillage or leakage of refined oils and the subsequent contamination of groundwater by the more water-soluble constituents — aromatic hydrocarbons such as benzene, toluene and xylenes. Pollution may be persistent and relatively high levels may be attained.

A class of hydrocarbons of particular concern is polyaromatic hydrocarbons (PAHs) as some PAHs are carcinogenic, for example, benzo[a]pyrene. The main source of these substances is incomplete combustion of fossil fuels. Possible human health risks, accumulation in marine sediments and effects on aquatic life have led to some concern. Levels in lowland river waters can be relatively high but since most of the loading is associated with suspended particulate matter, few water treatment problems appear to be generated [43]. However, marked increases in certain PAHs due to the bituminous linings of drinking water distribution mains have been noted [43]. Table 2.9 summarises levels of PAHs found in different types of water.

Table 2.9
Total PAHs in fresh and marine waters

Groundwater/upland water	ND – 50 ng/l	
Lowland rivers	40–300 ng/l	ref. [43]
Lowland rivers (high flows)	up to 3000 ng/l	
Marine sediments, polluted	up to 8500 ppb (dry wt. basis)	ref. [44]
Marine sediments, unpolluted	up to 97 ppb (dry wt. basis)	

Halogenated hydrocarbons

Halogenated hydrocarbons include a variety of important water pollutants due mainly to widespread usage, toxicity and in some cases an ability to accumulate in aquatic organisms.

Chlorinated hydrocarbon solvents, such as trichloroethylene, are commonly found in all types of natural waters but concentrations are usually $<1\mu g/l$. Surprisingly high levels of tri- and tetrachloroethylene ($>100\mu g/l$) have been measured in some groundwaters [45, 46]. Such contamination, which may be relatively widespread, presents some worrying aspects, especially when viewed in the light of proposed WHO drinking water guideline values for such compounds [47].

Contamination of water by halogenated hydrocarbons, for example chloroform, is possible due to *in-situ* synthesis by reaction of chlorine with organic matter, such as humic substances. Thus, chlorination of water supplies can lead to levels of chloroform, which is claimed to be a suspect carcinogen, above $100\mu g/l$ [48]. Other chlorinated organic substances, as yet largely unidentified, are also produced. Industrial chlorination, such as wood pulp bleaching, also leads to complex mixtures of chlorinated chemicals which can be detected in effluents and receiving waters [49].

Polychlorinated biphenyl (PCB) is a synthetic substance which has has widespread industrial application. Extreme persistence in the enviroment generally, an ability to accumulate in organisms and potential toxicity have led to considerable restrictions on its usage. The main source is wastewater and most PCB becomes associated with particulate matter.

Levels in drinking water are normally negligible. Levels in other types of water have been summarised [50] as follows:

freshwater	10–3000 ng/l
marine water	0.5–1600 ng/l (polluted coastal water).

Pesticides

Pesticides include a wide variety of substances and many find their way into the water cycle. The major entry into water is due to run-off from agricultural land and effluents from industries that use or manufacture them. Of the various pesticides, many chlorinated hydrocarbon insecticides are recognised pollutants, due in part to their persistence and ability to accumulate dramatically at certain stages of the food chain. The problems caused to aquatic and non-aquatic life are well-established. Studies on chlorinated hydrocarbon insecticides in water are numerous but the overall conclusions are as follows. Although they are frequently detected in drinking water, levels are invariably negligible. Levels in surface waters can vary significantly. Levels within the following ranges have been found in lowland river waters [51, 52]:

Dieldrin	not detected–60 ng/l
γ-BHC	1 ng/l–420 ng/l
α-BHC	not detected–570 ng/l
HCB	not detected–1200 ng/l
DDT (complex)	not detected–310 ng/l

Considerably higher concentrations have been recorded in situations where certain industrial effluents are discharged.

The presence of chlorinated hydrocarbons (insecticides and PCBs) in estuaries has been reviewed recently by Reutergardh [53]. He points out that these substances have very poor solubility in water. However, rivers contain humic substances and other organic material which behave as carriers for such substances. On mixing with sea water in estuaries the substances are 'salted out'. Thus, most studies have concentrated on levels in sediments and uptake by aquatic organisms. A conclusion from several studies is that the biotic residues of several chlorinated pesticides, for example DDT and dieldrin, have declined significantly over recent years.

2.4 ANALYTICAL CONSIDERATIONS AND GENERAL CONCLUSIONS

Water analysis is undertaken for one or more of a variety of reasons. Common objectives include:

(i) Monitoring to check compliance with standards.

(ii) Surveys of the concentration of specific pollutants or suspected pollutants to determine their distribution and potential environmental impact and monitoring of such substances to detect trends in time. Such analysis is often undertaken on a 'fire brigade' basis — that is after the existence of undesirable effects has become apparent. Examples of this are DDT and PCBs.
(iii) General surveys to determine the nature (and, though not invariably, the concentrations) of a wide range of potential pollutants, without prior evidence of undesirable effects related to particular substances.

Regrettably, much of the work to date on organic compounds in water has been of the 'fire brigade' type, although it is hoped (and anticipated) that growing experience, improved general survey techniques (item (iii) above) and the implementation of schemes to screen new chemicals for environmental effects will bring about a reduction in emphasis upon such work.

The ability to identify and, in many cases, to quantify a multitude of chemical impurities in water brings both benefits and problems. An important example is the use of gas chromatography—mass spectrometry. This technique produces data of undeniable value, and has shown the presence of substances at concentrations that are unacceptable — and of substances having no significance as pollutants. However, much of the data obtained falls into an intermediate category — that is, it is embarrassingly difficult to evaluate because the substances identified have not been the subject of suitable toxicological assessment.

This problem needs to be kept in perspective; it is relatively new and relates to environmental pollution in general. The problem has stimulated toxicological research and the development of suitable bioassay tests and therefore it seems likely that considerable advances will be made in this area. No doubt confidence in making practical judgements on the basis of incomplete information will increase in this field. One could decide not to undertake such survey work unless deleterious effects had been clearly disclosed. Such an attitude does not have much merit because one would forfeit advanced warning of potential problems and thereby forgo much of the benefit of modern analytical capabilities.

Application of such techniques must take into consideration their very limited availability — due mainly to costs. The techniques have been applied only relatively recently — within the last decade — and to a great extent only a general picture of the nature of pollution has been achieved. On the basis of this we anticipate more specific applications in the future in order to fill in important gaps in knowledge. An additional 'focussing' of techniques is their use in conjunction with the application of short-term bioassay tests in order to identify 'active' substances in water, for example mutagens.

Analysis, both for organic and inorganic substances, is subject to a number of other problems and limitations. The following are specifically noted.

(a) The accuracy of individual results, and the comparability of results obtained by different laboratories, are often poorer than is expected or tolerable. It is strongly recommended that an appropriate programme of Analytical Quality Control (AQC) should accompany any monitoring activity (for example, reference [54]); reliance upon 'standard' or 'recommended' analytical methods alone is insufficient. The need for AQC in water analysis is increasingly recognised and much progress has been made, but more remains to be done — particularly in respect of the determination of organic substances and of trace metals (especially in saline waters).

(b) Many pollutants are present, and of concern, at very low concentrations. This brings potentially severe problems, including the need to use highly sophisticated instrumentation and the very real dangers of contamination or loss of the determinand during analytical operations. Moreover, it must be clearly recognised that the ultimate accuracy of results depends upon the entire analytical method, not merely the instrumental 'finish'. Errors arising during necessary pretreatment stages, rather than those in the instrumental step, are likely often to determine the overall accuracy of analytical data.

(c) Powerful though modern analytical techniques undoubtedly are, there remain considerable gaps in our armoury and consequently in our knowledge. Of major concern are the difficulties of determining the non-volatile fraction of organic compounds, not amenable to GC–MS techniques, and of characterising and measuring the different trace metal (and metalloid) species. Although it has been suggested that the volatile fraction covers the majority of synthetic organic compounds or of important substances, there seems little basis to such statements and new methods for characterising non-volatile substances are needed. Some advances in this area have been made [55].

The potential benefits of an improved knowledge of trace element specification have already been described. New analytical methodologies are required for this purpose also.

In conclusion, we note that modern analytical techniques have provided us with greatly increased knowledge of the nature and concentration of pollutants and potential pollutants in water, though large gaps in our knowledge still remain. The power — and concomitant complexity — of such techniques bring their own problems, and these must be dealt with if the full benefits of the modern technology are to be realised. Specifically, the objectives of monitoring and surveillance schemes must be clearly defined, and the progress of such schemes regularly reviewed, to avoid wastage of analytical effort and appropriate procedures must be employed to ensure that the results are of adequate accuracy for the purpose(s) for which they are being obtained.

2.5 REFERENCES

[1] Council Directive of 4 May 1976 on pollution caused by certain dangerous substances discharged into the aquatic environment of the Community, 76/464/EEC, *Official Journal of the European Communities*, 1976, **19**, L129, pp. 23–29.

[2] Glover, H. G. (1975) Acid and Ferruginous mine drainage. *In* Chadwick, M. J. and Goodman, G. T., eds., *The Ecology of Resource Degradation and Renewal*, Blackwell, Oxford, pp. 173–195.

[3] Williams, S. L., Aulenbach, D. B. and Clesceri, N. L. (1974) Sources and Distribution of Trace Metals in Aquatic Environments. *In* Rubin, A. J., ed., *Aqueous-Environmental Chemistry of Metals*, Ann Arbor Science, Ann Arbor, (Michigan), pp. 77–127.

[4] Overrein, L. N., Seip, H. M. and Tollan, A. (1981) Acid Precipitation – Effects on Forrest and Fish, Final Report of the SNSF Project, 1972–1980, 2nd Edition, Oslo.

[5] D'Itri, F. M., ed., (1982) *Acid Precipitation. Effects on Ecological Systems*, Ann Arbor Science, Ann Arbor, (Michigan).

[6] Alabaster, J. S. and Lloyd, R. (Eds.) (1980) *Water Quality Criteria for Freshwater Fish*, Butterworths, London, 1980.

[7] Harriman, R. and Morrison, B. R. S. (1982), *Hydrobiologia*, **88**, 251–263.

[8] Galloway, J. N., Cosby, B. J., and Likens, G. E., (1979) *Limnology and Oceanography*, **24**, 1161–1165.

[9] Organisation for Economic Co-operation and Development, *Eutrophication of Waters. Monitoring, Assessment and Control*, OECD, Paris, 1982.

[10] Wilson, A. L. (1976) *Concentrations of Trace Metals in River Waters: A Review*. Water Research Centre Technical Report TR 16.

[11] Riley, J. P. and Chester, R. (Eds.) (1971) *Introduction to Marine Chemistry*. Academic Press, London.

[12] Batley, G. E. (1983) The current status of trace element speciation studies in natural waters. *In* Leppard, G. G., ed., *Trace Element Speciation in Surface Waters and its Ecological Implications*, Plenum Press, New York, pp. 17–36.

[13] Packham, R. F. (1983) Chapter 7 of this book.

[14] King, N. J., Hinchcliffe, P. R. and Vosser, J. L. Chapter 1 of this book.

[15] Stumm, W. and Bilinski, H. (1973) Trace metals in natural waters: difficulties of interpretation arising from our ignorance on their speciation, Jenkins, S. H., ed., *Advances in Water Pollution Research*, Proceedings of the 6th International Conference (Jerusalem, 18–23 June, 1972), Pergamon, Oxford, pp. 39–52.

[16] Florence, T. M. (1977) *Water Research*, **11**, 681–687.

[17] Gardiner, J. (1974) *Water Research*, **8**, 23–30.

[18] Anderson, D. M. and Morel, F. M. M. (1978) *Limnology and Oceanography*, **23**, 283–295.

[19] Gillespie, P. A. and Vaccaro, R. F. (1978) *Limnology and Oceanography*, **23**, 543–548.
[20] Hunt, D. T. E. and Creasey, J. D. (1980) *The Calculation of Equilibrium Trace Metal Speciation and Solubility in Aqueous Systems by a Computer Method, with Particular Reference to Lead*, Water Research Centre Technical Report TR 151.
[21] Patterson, J. W., Allen, H. E., and Scala, J. J. (1977) *Journal of the Water Pollution Control Federation*, **49**, 2397–2410.
[22] Jackson, P. J. and Sheiham, I. (1980) *Calculation of Lead Solubility in Water.* Water Research Centre Technical Report TR 152.
[23] Commins, B. T. and Packham, R. F. (1981) Population Exposure to Metals Released from Piping Materials used for Water Distribution in the UK. Commission of the European Communities' Second Environmental Research Programme 1976–80, Reports on Research Sponsored Under the Second Phase 1979–80, CEC, pp. 640–645.
[24] Directorate of Fisheries Research, Ministry of Agriculture, Fisheries and Food, *Atlas of the Seas around the British Isles*, MAFF, 1981, sections 2.22–2.23.
[25] Fenical, W. (1981) Natural halogenated organics. *In* Duursma, E. K. and Dawson, R., eds., *Marine Organic Chemistry*, Elsevier, Oxford, pp. 375–393.
[26] Commission of the European Communities. A Comprehensive List of Polluting Substances which have been Identified in Various Freshwaters, Effluent Discharges, Aqautic Animals and Plants, and Bottom Sediments. EURO-COST Project 64b, Commission of the European Communities, Brussels, 1983.
[27] Mackinnon, M. D. (1981) The measurement of organic carbon in sea-water. *In* Duursma, E. K. and Dawson, R., eds., *Marine Organic Chemistry*, Elsevier, Oxford, pp. 415–443.
[28] Cranwell, P. A., (1975) Environmental organic chemistry of rivers and lakes, both water and sediment. *Environmental Chemistry*, Vol. 1, The Chemical Society, London, pp. 22–54.
[29] Garland, J. H. N. (1978) Behaviour of degradable chemicals in the river Trent. *In* Hutzinger, O., Van Lelyveld, I. H., and Zoeteman, B. C. J., eds., *Aquatic Pollutants: Transformation and Biological Effects*, Pergamon, Oxford, pp. 275–281.
[30] Cauwet, G. (1981) Non-living particulate matter. *In* Duursma, E. K. and Dawson, R., eds., *Marine Organic Chemistry*, Elsevier, Oxford, pp. 71–89.
[31] Duursma, E. K. and Dawson, R., eds., (1981) *Marine Organic Chemistry* Elsevier, Oxford.
[32] *Environmental Chemistry*, Vol. 1, The Chemical Society, London, 1975.
[33] Hunter, K. A. and Liss, P. S. (1981) Organic sea surface films. *In* Duursma,

E. K. and Dawson, R., eds., *Marine Organic Chemistry*, Elsevier, Oxford, pp. 259–298.
[34] Waggot, A. (1976) *Analysis of the Organic Carbon Content of Sewage Effluent: General and Specific Group Analysis.* Water Research Centre Technical Report TR 29.
[35] Croll, B. T. (1972) *Water Treatment and Examination*, 21, 213–238.
[36] Bedding, N. D., McIntyre, A. E., Perry, R., and Lester, J. N. (1982) *The Science of the Total Environment*, 25, 143–167.
[37] Fielding, M., Gibson, T. M., James, H. A., McLoughlin, K., and Steel, C. P. (1981) *Organic Micropollutants in Drinking Water.* Water Research Centre Technical Report TR 159.
[38] Waggot, A. (1981), Trace organic substances in the River Lee. *In* Cooper, W. J. ed., *Chemistry in Water Re-use*, Vol 2, Ann Arbor Science, An Arbor (Mitchigan), pp. 55–99.
[39] Keith, L. H., ed., (1976) *Identification and Analysis of Organic Pollutants in Water*, Ann Arbor Science, Ann Arbor (Mitchigan).
[40] Keith, L. H., ed., (1981) *Advances in the Identification and Analysis of Organic Pollutants in Water*, Vols. 1 and 2, Ann Arbor Science, Ann Arbor (Mitchigan).
[41] Gschwend, P. M., Zafiriou, O. C., Mantoura, R. F. C., Schwarzenbach, R. P., and Gagosian, R. B., (1982) *Environmental Science and Technology*, 16, 31–38.
[42] Farrington, J. W. and Meyer, P. A., (1975) Hydrocarbons in the environment. *In Environmental Chemistry*, Vol 1, The Chemical Society, London, pp. 105–136.
[43] Crane, R. I., Fielding, M., Gibson, T. M., and Steel, C. P., (1981) *A Survey of Polycyclic Aromatic Hydrocarbon Levels in British Waters.* Water Research Centre Technical Report TR 158.
[44] Hites, R. A., Laflarmine, R. E., and Windsor, J. G. (1980) Polycyclic aromatic hydrocarbons in the marine environment. *In* Afghan, B. K. and Mackay, D., eds., *Hydrocarbons and Halogenated Hydrocarbons in the Aquatic Environment*, Plenum, London, pp. 397–403.
[45] US Environmental Protection Agency, (1982) *Federal Register*, United States Environmental Protection Agency, Washington, March 4, 47, 9350–9358.
[46] Fielding, M., James, H. A., and Gibson, T. M. (1981) *Environmental Technology Letters.* 2, 545–550.
[47] Gorchef, H. G. and Ozolins, G., (1982) *WHO Guidelines for Drinking Water Quality.* Paper presented at the International Water Supply Association Congress, September, Zurich, International Water Supply Association, London.
[48] *Trihalomethanes in Water.* Proceedings of a Water Research Centre Seminar, Water Research Centre, 1980.

[49] Lindström, K., Nordin, J., and Österberg, F. (1981) Chlorinated organics of low and high relative molecular mass in pulp mill bleachery effluents. *In* Keith, L. H., ed., *Advances in the Identification and Analysis of Organic Pollutants in Water*, Vol 2, Ann Arbor Science, Ann Arbor (Michigan), pp. 1039–1058.

[50] IARC Monographs in the Evaluation of the Carcinogenic Risk of Chemicals to Humans – Polychlorinated Biphenyls and Polybrominated Biphenyls, Vol. 18, International Agency for Research in Cancer, Lyon, 1978.

[51] Croll, B. T. (1969) *Water Treatment and Examination*, 18, 255–274.

[52] Wegman, R. C. C. and Greve, P. A. (1980) Halogenated hydrocarbons in Dutch water samples over the years 1969–1977. *In* Afghan, B. K. and Mackay, D., eds., *Hydrocarbons and Halogenated Hydrocarbons in the Aquatic Environment*, Plenum, London, pp. 405–415.

[53] Reutergarddh, L. (1980) Chlorinated hydrocarbons in estuaries. *In* Olausson, E. and Cato, O., eds., *Chemistry and Biogeochemistry of Estuaries*, Wiley, Chichester, pp. 349–365.

[54] Wilson, A. L., (1979) *Analyst*, 104, 273–289.

[55] Watts, C. D., Crathorne, B., Fielding, M., and Steel, C. P. (1983) *Identification of Non-Volatile Organics in Water using Field Desorption Mass Spectrometry and High Performance Liquid Chromatography*. Paper presented at the Third European Symposium on Analysis of Organic Micropollutants in Water, Oslo, 19–21 September.

CHAPTER 3

Pollution – the risk in perspective

SIR FREDERICK WARNER, FRS
Cremer and Warner

The placing of risks in perspective has to follow the process of risk assessment, which is the study of decisions subject to uncertain consequences. It is divided into risk-estimation and risk-evaluation. The former includes:

(a) the identification of the outcomes;
(b) the estimation of the magnitude of the associated consequences of these outcomes;
(c) the estimation of the probabilities of these outcomes.

Risk-evaluation is the complex process of determining the significance or value of the identified hazards and estimated risks to those concerned with or affected by the decision. It therefore includes the study of risk perception and the trade-off between perceived risks and perceived benefits. Risk-management is the making of decisions concerning risks and their subsequent implementation, and flows from risk-estimation and risk-evaluation.

The definition of risk for the purpose of this chapter is taken from the Royal Society Group Report [1] published in 1983. Risk is 'the probability that a particular adverse event occurs during a stated period of time, or results from a particular challenge'. As a probability in the sense of statistical theory risk obeys all the formal laws of combining probabilities. The techniques of risk assessment go back to the need for reliability in military equipment and air transport. They now use the data banks which hold increasing amounts of information to provide failure rates. The whole subject is becoming of increasing importance in the search to provide reliable goods at a competitive price. It is at the centre of a major Government strategy to improve industrial performance by encouraging quality assurance procedures. These are covered in a number of British Standards included in Handbook No. 22. The glossary of terms is to be found in BS 4778.

The water industry has to set its own standards of quality which are based upon chemical, bacteriological, and organoleptic criteria. They have as their objective to ensure a supply of water to households and industry which can be drunk for a lifetime without giving rise to serious risk to health or life. They have also to ensure that rivers, estuaries and the sea are not polluted so that there is a reduction in genetic variety, the loss of species or the encouragement of out-of-balance growth. Nor must the aesthetic value be overlooked. It is taken for granted that the procedures will ensure that sudden failures, resulting in a crisis, can be prevented — that there can never be a sudden danger to human health. These procedures will protect a public water supply but cannot be totally effective against accidents or wilful damage which lead to pollution of rivers by oil, caustic soda, sulphuric acid or any of the materials in common use which are transported or stored in quantity.

These are the considerations in a developed country which takes for granted that all but the most remote households can have a supply of piped water and facilities for water-borne sewage disposal and treatment. For less developed countries, the picture is quite different and the number of deaths related to water-borne diseases in the world is put at something around 20 million a year. It is in this area that the efforts of international agencies such as WHO will be directed, following their success over the last fifteen years in completely eliminating 2 million deaths a year from smallpox.

To set the risks of pollution in perspective is to face the problem that the information available on risk is limited. Even in the countries which have statistics over a long period, as in Great Britain and the USA, these are overwhelmingly related to mortality and not to sickness or indeed fates worse than death. In a 1978 publication, Grist [2] brought together the most recent British data available in individual risk and used the International Classification of Diseases (ICD) to group them under various heads. Table 3.1 shows the individual risk per year for various age groups. Row 1 covers all the 1000 ICD categories, Row 2 all natural causes, and Row 8 all violent causes of death, averaged over five years. Table 3.2 shows a breakdown of risk estimates for accidental death in different forms. About one-third of accidental deaths are due to road accidents and all deaths due to accidents and violence (excluding Northern Ireland) are shown in Table 3.3. Of road accidents, the greatest number of deaths occur in the 15 to 19 age group, followed by the 65 to 74 group. It remains to be seen how far deaths on the road are reduced from 1 January 1983 with the compulsory wearing of seat-belts in cars.

In estimating risk, the attempt is made to use *in vitro* or animal experiments to indicate an effect and extrapolate to man by the use of a model. They are of limited value and do not elucidate the reality of hazards to health such as are provided by epidemiological observations on man. The Royal Society Report gives five main reasons. First, some hazards arise from behaviour that is uniquely human (car driving, mountaineering, homicide, deep-sea fishing). Second, a

Pollution − The Risk in Perspective

model may be based on incomplete knowledge of the hazards. Laboratory experiments did not, and probably could not, predict the effects on membranes of prolactol, pulmonary hypertension from the appetite-depressant 'aminorex', lung cancer from arsenic or leukaemia from benzene. Third, biological models are limited in the prediction of dose−effect relationships. Observations on man can demonstrate the size of risk associated with a given exposure as for example to ionising radiation. This can then help to determine what exposure (if any) is socially acceptable in view of the benefits to society as a whole. A new medicine may be shown in laboratory experiments to have carcinogenic effects, but its beneficial effects make it worthwhile to expose humans and diagnose adverse effects by the normal medical service of the country. The results are more reliable than the largest experiments on animals. The continual monitoring after introduction of new chemicals, machines or processes is the fourth reason to check that the risk is contained within approved limits. The fifth, and most important reason is that human observations provide a sense of perspective.

The general background for the study of risk has been set and the particular problems posed by pollution remain. Those which have been the subject of publicity and comment over the years, not in any particular order are:

Detergents
Phosphates
Borates
Nitrates
Halogens and compounds
Lead, mercury, zinc and cadmium
Pathogens including viruses
Asbestos fibres
Radioactive agents
Hormones

The polluting effects of detergents, in so far as the main surface-active components are concerned, are now a matter of history. They were in any case a secondary effect, causing problems in sewage treatment works by reducing the efficiency of operations by the interference of foam in the purification mechanisms. The story of substitution of the original 'hard' (non-biodegradable) detergents by biodegradable materials has been completely described. The reports of the Standing Technical Committee on Detergents (of which the author was chairman for a period) began in its early work by the identification in the Metropolitan Water Board laboratories. It might be said that this outstanding work in the field of chemical analysis ironically led to the problems which have been posed from the increasing ability to detect chemicals in concentrations as low as one in a billion (10^{-9}).

Table

Risk of death as a function

ICD Nos.	Row number	Cause of death	All ages	0–4	5–9
000–999	1	All causes	11910	3693	334
000–799	2	Natural causes	11453	3427	203
140–239	3	Neoplasms	2475	79	68
390–458	4	Diseases of the circulatory system	6177	27	7
410–414	5	Ischaemic heart disease	3120	‡	‡
480–486	6	Pneumonia	887	320	16
520–577	7	Diseases of the digestive system	296	67	6
800–999	8	All deaths due to accidents and violence	457	267	131
950–959	9	Suicides	78	No deaths recorded	
960–999	10	Homicides, legal, undetermined, war	36	26	6
800–949	11	All accidents	343	240	125
810–823	12	Road accidents	138	54	78
846–949	13	All non-transport accidents	196	184	44
% of all deaths due to accidents		$\dfrac{(800-949)}{(000-999)} \times 100$	2.88	6.49	37.43

† Fewer than 50 but more than 20 deaths registered in the period 1971–1975.
‡ Fewer than 20 deaths registered in 1971–1975.
Data presented were obtained from mortality and total population statistics for Great Britain.

3.1

of age. 1975-1971 mean.

			Individual risk per year $\times 10^6$						
10-14	15-19	20-24	25-34	35-44	45-54	55-64	65-74	75-84	85+
282	637	718	791	1945	5897	15261	37642	90023	207405
174	250	310	483	1622	5504	14772	36934	88188	202378
55	71	95	171	583	2005	4983	9505	14143	17025
15	27	45	121	654	2535	7180	19879	52558	127752
‡	‡	4	39	391	1711	4655	11278	23617	45983
13	21	21	22	51	142	466	1976	8631	29720
7	11	17	26	69	178	391	926	2239	4522
108	387	408	308	324	393	489	708	1835	5027
1†	25	64	76	96	125	138	150	138	94
5	27	38	33	41	48	52	51	57	54
101	336	306	199	186	219	229	506	1639	4879
57	259	215	117	92	98	139	203	363	340
39	65	78	70	84	111	150	297	1268	4530
35.79	52.69	42.67	25.17	9.56	3.71	1.96	1.34	1.82	2.35

Table 3.2
Individual risk estimates (accidents and violence)

ICD	Description of cause	Individual risk (per year, $\times 10^6$)	Notes	0.95 probability confidence interval
800–807	Railway accidents	3.3	(b)	
810–823	Motor vehicle traffic accidents and motor vehicle non-traffic accidents	122	(a)	
814	Motor vehicle traffic accidents involving collision with pedestrian	45	(a)	
830–838	Water transport accidents	2.9	(b)	
840–845	Air and space transport accidents	1.4	(b)	
842	Accident to unpowered aircraft	7.4×10^{-2}	(c)	$5.0 \times 10^{-2} - 10.4 \times 10^{-2}$
850–859	Accidental poisoning by drugs and medicaments	11	(a)	
855–856	Accidental poisoning by other drugs	1.9	(a)	
800–869	Accidental poisoning by other liquid and solid substances	2.2	(a)	
868	Accidental poisoning by noxious foodstuffs and poisonous plants	9.2×10^{-3}	(c)	$2.3 \times 10^{-3} - 23.5 \times 10^{-3}$
870–877	Accidental poisoning by gases and vapours	3.2	(a)	
880–887	Accidental falls	111	(a)	
881	Falls on or from ladders or scaffolding	2.1	(a)	
890–899	Accidents caused by fire and flames	16	(a)	
900–909	Accidents due to natural and environmental factors	4.2	(c)	
900	Excessive heat	4.6×10^{-2}	(d)	$3.4 \times 10^{-2} - 6.2 \times 10^{-2}$
901	Excessive cold	1.2	(c)	
904	Hunger, thirst, exposure and neglect	2.3	(c)	
905	Bites and stings of venomous animals and insects	8.5×10^{-2}	(d)	
906	Other accidents caused by animals	4.0×10^{-1}	(d)	
907	Lightning	1.0×10^{-1}	(d)	

Table 3.2 – *continued*
Individual risk estimates (accidents and violence)

ICD	Description of cause	Individual risk (per year, $\times 10^6$)	Notes	0.95 probability confidence interval
908	Cataclysm	4.3×10^{-2}	(d)	3.1×10^{-2} – 5.8×10^{-2}
909	Accident due to other natural and environmental factors	9.2×10^{-3}	(c)	2.3×10^{-3} – 23.5×10^{-3}
910–929	Other accidents	43	(a)	
910	Accidental drowning and submersion	12	(a)	
911	Inhalation and ingestion of food causing obstruction or suffocation	11	(a)	
916	Hit by falling object	3.4	(b)	
922	Accident caused by firearm missiles	6.2×10^{-1}	(c)	
923	Accident caused by explosive material	8.3×10^{-1}	(c)	
925	Accident caused by electric current	2.5	(b)	
930–936	Surgical and medical complications and misadventures	1.5	(b)	
940–949	Late effects of accidental injury	2.2	(b)	
950–959	Suicide and self-inflicted injury	76	(a)	
960–969	Homicide and injury purposely inflicted by other persons	10	(a)	
980–989	Injury undetermined whether accidentally or purposely inflicted	30	(a)	
800–999	All deaths due to violence	442	(a)	
800–949	Deaths from all accidents	325	(a)	
850–949	Deaths from all non-transport accidents	195	(a)	

(a) Based on 1975 data.
(b) Based on 1975–71 data.
(c) Based on 1975–1969 data.
(d) Based on 1975–1958 data.
0.95 probability confidence intervals are quoted where the risk calculation employed a total of less than 50 deaths recorded in the appropriate time period.

Table

Deaths due to accidents and

ICD	Description	1975	1974	1973	1972	1971	1970
800–807	Railway accidents	181	180	172	142	212	209
810–823	Motor vehicle traffic accidents and motor vehicle non-traffic accidents	6651	7219	7847	7797	7896	
814	Motor vehicle traffic accidents involving collision with pedestrian	2441	2798	2921	3124	3033	
830–838	Water transport accidents	171	141	143	171	170	173
840–845	Air and space transport accidents	41	57	181	55	49	46
842	Accident to unpowered aircraft	5	8	5	1	2	3
850–859	Accidental poisoning by drugs and medicaments	601	591	545	539	581	
855–856	Accidental poisoning by drugs (nervous system)	105	106	83	64	49	
860–869	Accidental poisoning by other liquid and solid substances	120	103	85	81	82	65
868	Accidental poisoning by noxious foodstuffs and poisonous plants	0	2	0	0	0	1
870–877	Accidental poisoning by gases and vapours	172	179	214	231	318	
880–887	Accidental falls	6019	6114	6014	6070	6235	6316
881	Fall on or from ladder or scaffolding	113	121	112	113	122	109
890–899	Accidents caused by fire and flames	846	909	968	892	803	931
900–909	Accidents due to natural and environmental factors	250	216	192	217	209	258
900	Excessive heat	7	1	2	1	1	3
901	Excessive cold	42	38	38	55	65	106
904	Hunger, thirst, exposure and neglect	163	148	112	136	97	106
905	Bites and stings of venomous animals and insects	3	1	3	0	6	7
906	Other accidents caused by animals	35	23	30	19	34	23
907	Lightning	5	1	5	4	4	11
908	Cataclysm	0	0	1	0	0	0
909	Accident due to other natural and environmental factors	0	1	0	0	0	1
910–929	Other accidents	2347	2351	2318	2348	2616	2669
910	Accidental drowning and submersion	629	499	581	548	618	608
911	Inhalation and ingestion of food causing obstruction or suffocation	614	660	611	602	636	668
916	Struck accidentally by falling object	169	190	179	173	224	195
922	Accident caused by firearm missiles	28	34	34	38	35	34
923	Accident caused by explosive material	53	43	25	43	58	49
925	Accident caused by electrical current	108	141	134	146	143	151
930–936	Surgical and medical complications and misadventures	101	86	89	64	57	
940–949	Late effects of accidental injury	130	119	117	117	103	
950–959	Suicide and self-inflicted injury	4120	4336	4259	4191	4323	4337
960–969	Homicide and injury purposely inflicted by other persons	565	547	519	457	473	398
980–989	Injury undetermined whether accidentally or purposely inflicted	1638	1378	1327	1340	1367	1346
800–999	All deaths due to accidents and violence	24045	24606	25074	24801	25582	
800–949	Deaths from all accidents	17704	18335	18947	18803	19406	
850–949	Deaths from non-transport accidents	10585	10668	10542	10559	11005	

† Denotes that although earlier data is really needed to ensure statistical significance pre-1968 data is not

3.3

violence (Great Britain)

1969	1968	1967	1966	1965	1964	1963	1962	1961	1960	1959	1958
202	150										
194	158										
107	147	124	73	99	69	43	83	44	63	81	204
2	6†										
50	51										
1	0†										
6455	6141										
132	123										
878	956										
241	237										
4	4	0	1	1	5	3	0	2	4	5	0
115	39†										
101	143†										
4	1	7	4	5	2	9	5	7	4	4	9
11	23	13	11	22	25	14	21	22	15	20	22
3	8	6	1	1	3	2	5	9	9	7	12
0	23	0	1	1	1	2	8	0	1	2	1
2	0†										
2639	2560										
607	527										
667	660										
240	198										
27	38										
40	50										
143	135										
688	4957										
395	404										
670	1440										

mpatible with the ICD 8th revision.

The phosphate story occupied the front of the stage for many years because of eutrophication in lakes and the sudden appearance of 'blooms' of algae and other growths. The substitution of phosphates by sodium nitrilotriacetate was advocated without serious assessment of the effects of such a new compound in place of one which had been lived with for a long time. Phosphates are not even mentioned in the index of the seventh report of the Royal Commission on Environmental Pollution 'Agriculture and Pollution' [3]. Of boron arising from use of perborates for bleaching action in detergents nothing is heard since the cessation of the reports of the Standing Technical Committee.

Nitrates have by contrast come to the fore in discussion of risks from drinking water. The Royal Commission concluded:

> We do not doubt that the large increase that has occurred in the use of inorganic nitrogen fertilizer is, directly or indirectly, a major cause of the rising nitrate levels that are observed in many water supply sources. There is concern about these levels because of the risk that nitrate may cause methaemoglobinaemia in infants or that it may be implicated in some forms of human cancer through the formation of N-nitroso compounds, which are known to be powerful carcinogens in a variety of animal species. We have therefore regarded it as an important part of our study to assess these risks, the contribution to them from agricultural practices, and the remedial measures that might be taken should concerted action to reduce nitrate levels appear essential.
>
> The health risks posed by nitrate in water supplies have attracted much public comment; public concern on the matter has been augmented by the possibility that nitrate levels in groundwater have yet fully to reflect past increases in fertilizer use, so that these levels may continue to rise. We have reached the conclusion, however, that the anxiety that has been engendered about these risks is not justified on the information at present available. Infant methaemoglobinaemia is a rare condition in the UK, even in the few areas where water with nitrate content above the 'recommended' level of 50 mg/l (expressed as nitrate) in the WHO European standards has been consistently supplied. With regard to the possible cancer risk, the link with nitrate intake is highly uncertain and we have found no evidence that water containing 100 mg nitrate/l is harmful to adults. We see no basis for departing from the WHO European standards in the UK; that is, we think the 50 mg/l limit is prudent for infants but that 100 mg/l should be taken as the maximum acceptable level for the present. We recommend, however, that further investigation is undertaken to substantiate this maximum acceptable level, particularly in relation to the risk of infant methaemoglobinaemia under UK conditions.
>
> Although agricultural practice certainly affects nitrate levels in water, the connection between them is complex; in particular, there is much

Pollution – The Risk in Perspective

certainty about the effect that a reduction in fertilizer use would have in reducing these levels.

On the question of cancer and nitrate in the diet, including water, Doll and Peto [4] conclude 'there is thus little evidence that dietary nitrates, nitrites, or nitrosable compounds contribute to the production of gastric cancer or any other type of cancer, but this may merely be because there is inadequate evidence on the whole subject'.

Halogens and their compounds in the form of insecticide and herbicide formulations have received most examination and comment ever since Rachel Carson wrote *Silent Spring* in 1962. The bio-accumulation of chlorinated hydrocarbons and their effect on wild-life is commented on in the Royal Commission Report. As regards human targets, Doll and Peto say under *Drinking Water*.

The situation with regard to pollution of drinking water is rather more obscure. Analytical techniques now permit the detection of chemicals at concentrations well below 1 ppb (and further rapid improvements in sensitivity must be anticipated), and consequently, many chemicals have been found to contaminate drinking water supplies. Some of these may be classed on the basis of laboratory tests as carcinogens or mutagens (or suspected carcinogens or mutagens), and the possibility that they may constitute a material hazard to health has been examined by the National Academy of Sciences (1977) [5]. As a result of its report, the EPA has promulgated a regulation establishing a 'maximum contaminant level' of $100\mu g$ total trihalomethanes/litre for all community water systems serving 10,000 persons or more and adding a disinfectant in the treatment of their water supplies.

The EPA did not claim that drinking water, as now supplied in the United States, had been proved to cause cancer. However, human exposure to water pollutants may be prolonged over so many decades that even weak carcinogenic stimulii might in principle have material effects, and the evidence that they might do so includes:

(a) There is activity in short-term carcinogenicity tests or animal cancer tests of certain pollutants (or, equivalently, there is an excess of tumours in marine animals that live in polluted waters).

(b) Some pollutants are present that are known or believed to cause cancer in humans if large amounts are ingested (e.g. asbestos from industrial activity, from asbestos cement water pipes, or from the passage of groundwater through natural asbestos-rich rock formations; radionuclides from industrial activity or natural sources; arsenic; and vinyl chloride from PVC water pipes).

(c) There are human population studies that report positive correlations between the amounts of certain contaminants and mortality from certain cancers.

After disposing of (b), particularly in respect of asbestos fibres, they conclude that from epidemiology

> The interpretation of the correlation studies is at present open to question. Similar studies have been carried out over the years in many other fields (for example, showing that 'Contamination' of drinking water by calcium appears to decrease the risk of heart disease: Crawford et al., 1968 [6]) but have rarely been regarded as constituting anything more than hypotheses to be tested by more specific work, because of the difficulty of obtaining truly relevant data (relating to long-past exposure of the actual individuals concerned) and of eliminating the effects of concomitant variation. (The specific difficulties inherent in correlation studies of cancer and drinking water composition have been emphasized by Hogan et al., 1979 [7].) Analyses that took account of other important variables and were consistent from one region to another in pointing to specific effects on one or other specific type of cancer would carry some weight, but most of the analyses have not met these criteria.

They go on to say that the chief exceptions are studies

> suggesting a relationship between the concentrations of halogenated organic matter in water and mortality from cancers of the bladder and, possibly, the large intestine. When water is chlorinated to kill germs, much of the plant debris and other organic matter naturally present in it is also halogenated, resulting in a complex mixture of halogenated compounds at levels of up to some parts per million (depending principally on the concentration of organic matter that was originally present). Many specific halogenated compounds have been found to cause tumours in laboratory animals and some have also done so in man. It is therefore reasonable to inquire whether lifelong exposure to the non-specific mixture present in water that has been chlorinated has any material effect on the risks form any type of tumour. If the large NCI case-control study of bladder cancer that is currently being analysed indicates any substantial relationship among individuals with the concentration of such compounds in drinking water (or if any really well controlled studies of cancer of the large intestine do so), then this will lend urgency to similar studies of many other types of tumour. However, in view of our previous estimate of some thousands of cancers which urban air pollution may cause by enhancing the carcinogenicity of cigarettes, it seems on present evidence that any effects of drinking water pollution will probably be relatively less important. If this is so, then they will not materially affect the total percentage of cancer deaths to be attributed to 'pollution'. But it has to be borne in mind that the long duration of exposure to any traces of carcinogens that drinking water may contain will enhance any effects they might have.

Their conclusions are that it is difficult to set an upper limit until the role, if any, of halogenated compounds in drinking water is clarified.

On fluorides in drinking water, they reject the claim that this increases the cancer risk. Since their study is confined to this, they do not commment on dental fluorosis associated with naturally high fluoride content of water in such places as Maldon and Nakuru which does not fall into the definition of pollution anyway.

Of the heavy metals, only lead is of importance in relation to water supplies. It has been so exhaustively discussed recently in relation to the source of the total body burden and the control of lead in petrol that no repetition is needed here. There has been no claim that deaths have resulted from ingestion of lead, but that behavioural changes are related to blood lead-levels. It is an area of risk which is difficult to enter because the statistics do not have the certainty of mortality figures. These do show the protective effect of calcium and magnesium in a number of cases and the arguments continue about the higher mortality from such things as ischaemic heart disease in soft water districts.

Radioactivity in water is carefully supervised by the National Radiological Protection Board in inland, estuarial and sea water while the Ministry of Agriculture, Fisheries and Food monitors bio-accumulation in fish, shellfish and seaweed. By concentration of effort on critical groups, they are able to ensure by monitoring that the amount of radioactivity taken in by the most exposed member of a critical group is below the permissible limit set by the International Commission on Radiological Protection.

Some long time ago the Water Research Centre at Medmenham monitored drinking water from the Thames for residual hormones associated with the spreading use of oral contraceptives. Many studies have been, and are being made of the direct effects on women on using the pill. The conclusion seems to be that the effect on production of cancers is small either way.

On pathogens and viruses, there is nothing useful to say. The whole industry is centred around the provision of safe drinking water. The protection of water generally from the discharge of sewage, whether treated wholly, partially, or untreated, is monitored so as to avoid the transmission of disease. It is no part of this chapter to enter into the controversy over sea-outfalls, EEC directives and a vague worry that there are sub-clinical effects now on a larger scale than there used to be.

It is also no part of this chapter to examine how far the expenditure on reducing water pollution has been cost-effective. If the spectacular example of the Thames estuary is considered, the benefits seem to be mainly aesthetic and ecological rather than those which can be quantified in terms of disease-prevention or fewer problems in industrial use.

It is ironical that the work instigated by Lord Waverley as Chairman of the PLA was inspired by the blackening of ships in the Pool of London. There is now no lead paint to blacken by sulphide, very few ships in the Pool either, and no dock employment. The risks that follow social and economic change seem to have much more far-reaching effects than any which can be laid at the door of pollution.

REFERENCES

[1] *Risk Assessment* (1983) ed. Sir Frederick Warner. Royal Society.
[2] Grist, D. R. (1978) *Individual Risk.* UKAEA Report SRD R125. HMSO.
[3] Royal Commission on Environmental Pollution (1979) Seventh Report. *Agriculture and Pollution.* HMSO.
[4] Doll, Sir Richard, and Peto, R. (1981) *The Causes of Cancer*, Oxford University Press.
[5] National Academy of Sciences (1977). *Drinking Water & Health*.
[6] Crawford, M. D., Gardner, M. J., Morris, J. N. (1968) Mortality and hardness of water supplies. *Lancet*, 1, 827–831.
[7] Hogan, M. D., Chi, P.-Y., Hoel, D. G., Mitchell, T. J. (1979) *J. Environ. Toxicol. Pathol.*, 2, 873–877.

Discussion

Chairman: **PROFESSOR T. R. E. SOUTHWOOD, FRS**
Chairman, Royal Commission on
Environmental Pollution

The discussion began with a description of river pollution in Cornwall caused by inert suspended solids from mine workings. In the case of china clay mining, 'white rivers' were caused by effluents from china clay pits. These now have largely disappeared as a result of effluent control measures begun in 1970. In another case effluents from new and old tin mines produce gross discoloration and the example of the Red River near St. Ives was shown.

The alleged relationship between increased fertiliser usage and higher levels of nitrate in groundwater, and the role of urban run-off in causing pollution, described in Dr King's chapter, were thought to be examples of where a great deal of excellent research had been carried out. However, there was a need to review: (a) the mechanisms, (b) the effects, and (c) the control options pertaining to these sources.

Dr N. J. King agreed with this assessment but pointed out that control of pollution was not within the remit of his chapter, and that the official position regarding the role of fertilisers in causing groundwater pollution was unresolved.

The speakers were asked that when considering the different classifications of pollutants was it not important to take account of the different biological pathways involved and the changes in 'speciation' and toxicity resulting from movement along these pathways?

Dr D. T. E. Hunt replied that research on the influence of chemical speciation on toxicity is hampered in two directions. Firstly it is difficult to characterise metal speciation either by experiment or analysis, and our ability to distinguish using thermodynamic calculation is also poor. Secondly, in order to determine differences in the toxic reactions of a test organism it is often necessary to alter other characteristics of the water quality. This change may produce reactions in the test organisms which would be difficult to separate from those produced by the changes in metal speciation.

Mr M. Fielding observed that our knowledge of the transport of organic pollutants through the environment is rather poor and, often, we discover the pathways only following an accident. This is also true of the mechanisms by which transport processes affect the form inorganic and organic pollutants take. With organic pollutants this lack of understanding means that research is often effected only when a problem is perceived.

The general responses of the speakers so far indicated that there are substances in the environment which might cause damage to health but that the risk is probably very small indeed. The question was raised; should the public at large be educated in the concept of acceptable risk and if so, how this should be achieved?

Sir Frederick Warner noted that one of the problems in risk perception is that one eventually comes down to ethical decisions. One has to set the cost of saving lives or preventing disease against the values we place on our own lives and then make the decision as to where we allocate the resources. In this context there is no general solution because it is a problem of individual ethics.

Mr P. Powell continued the theme of risk perception. He argued that voluntary risks (for example, cigarette smoking) are more acceptable to the public than involuntary risks (for example, carcinogens in food). This is dealt with in greater detail in the Support Paper, *The Perception of Risk*, by P. Powell and R. F. Lacey, in this volume.

Sir Frederick Warner responded by saying that the perception of risk was an area receiving great attention and he anticipated many papers in this field but, in general, people always expect to see a sequel to a given action. This does not always occur and this 'logical positivist' approach may lead to confusion over perception.

Referring to the first chapter, the point was raised that the design of the older, short, sea outfalls was inadequate and that up to 95% of them may lead to increased local pollution around the point of discharge. Implementation over the next four years of part 2 of the Control Of Pollution Act (COPA 2) might not resolve this problem because all existing discharges will be exempt from control unless they contain significant quantities of list 1 and list 2 substances. If they do contain these substances they will be given 'deemed consents' which will legalise their discharges at the existing level. Until there is a big improvement in the economic climate there may not be any change.

The need to look at processes involved in pathways of pollutants in particular, physical chemistry, biochemistry and environmental processes was raised. This received broad agreement but decisions sometimes have to be made with less than perfect data. One of the functions of managers is to decide when we have enough knowledge to make a reasonable decision.

Sir Frederick Warner agreed and made the case for the trial use of apparently safe chemicals on man because this provides the best model and best experiment which cannot be performed on any other animal.

Part 2

ASSESSMENT OF EFFECTS OF POLLUTION

CHAPTER 4

The role of toxicology

J. K. FAWELL
Water Research Centre

4.1 INTRODUCTION

Toxicology is the science which studies the adverse effects of chemicals on living organisms and assesses the probability of the occurrence of these effects. The assessment of the probability of occurrence of toxic effects is a relatively new field and one which presents particular difficulty for the toxicologist.

Toxicology is crucial in providing information in support of many facets of environmental protection. Wherever an interaction between a chemical and a living organism may produce adverse effects, information on toxicity becomes necessary. The requirement for toxicological information when investigating long- and short-term problems, both in operational and the more speculative research areas in the water industry, provide good examples of this interaction. Water is not only vulnerable to spills and accidental contamination from industrial and agricultural chemicals, but it is arguably the most convenient disposal route we have for many wastes. Almost any chemical may find its way into water and may cause a short-term operational problem if in sufficient quantities such as in the case of a chemical spill. In addition, many compounds are present only in small quantities but exposure to these may be for very long periods of time. Such is the situation with organic micropollutants. In this case the requirement is to determine whether a problem exists and if it is of sufficient magnitude to warrant a change in operational practice. Further, as the potential for exposure of most of the population through water is enormous, a great responsibility rests with the industry to ensure their own actions do not create new problems while solving others. Therefore, when examining the safety of materials and chemicals used in water by the industry, there is a change in approach from defensive to preventive. This philosophy is also apparent when standards or guidelines for contaminants in water are being developed.

All these considerations have enormous cost implications for the water industry and society in general so decisions cannot be taken lightly.

Toxicology is very much an applied science requiring a mixture of scientific disciplines; medicine, biology, pharmacology, biochemistry and chemistry all having a place in toxicology. Consequently toxicology is dependent to a great extent on the state of knowledge in these disciplines. However, the questions posed by toxicology are now providing their own stimulus for fundamental research.

The principles of all kinds of toxicology are basically the same but there are differences in approach throughout the science. In mammalian toxicology, for example, a significant effect is measured on the individual whereas in ecotoxicology a significant effect is gauged by assessing the stability of a population. In this chapter, comments will be restricted to mammalian toxicology. One of the particular interests and difficulties of environmental toxicology, is that we deal with the whole human population. Unlike our colleagues in, for example, the pharmaceutical industry, we cannot work with a limited target population. We must consider all ages and all states of health from the oldest to the youngest, from the healthy to the sick. It is therefore important, when examining toxicological data for the purpose of environmental protection, that the problem be placed in its correct context.

4.2 WHAT INFORMATION IS NEEDED?

There are a number of aspects of the biological activity of a compound on which information will be required. A range of studies can be used to generate such information on acute, subacute and chronic toxicity, and reversible and irreversible effects (Table 4.1). It should be noted that acute and chronic refer to the duration of exposure and not the severity of the effect. Acute toxicity has been defined as 'the total adverse effects produced in the organism when administered as a single dose or in multiple doses over a period of twenty-four hours or less' though these effects may occur some time after the dose [1]. Unfortunately acute toxicity has tended to become synonymous with the dose lethal to 50% of a group of test animals (LD50) but as many stages of injury occur before death, the LD50 is not a particularly helpful number in this respect.

One of the measures most frequently encountered in experimental toxicology is the dose response curve, a relationship first recognised by Paracelsus in the 16th century [2]. There are three interacting components to any assessment of dose response curves (Fig. 4.1) as follows:

1. The quantity of toxicant.
2. The time of exposure.
3. The severity of the effect.

Table 4.1
Aspects of biological activity and methods of study

1.	Acute toxicity	Acute LD50
		Sub-acute 30-day study
2.	Chronic toxicity	90-day study
		1 year plus study
3.	Carcinogenicity	Lifespan study
		Multigeneration study
		Epidemiology (occupational)
4.	Mutagenicity	Short-term tests: bacteria, yeasts mammalian cells
		In vivo studies: Drosophila, micronucleus test
5.	Reproductive toxicity	Teratology and embryotoxicity studies
		Fertility studies
		Multigeneration studies
6.	Metabolism and pharmacokinetics (toxicokinetics)	Radiotracer studies
		Analytical studies
7.	Special studies	Behaviour studies
		Immunotoxicity
		Allergy

All toxicity studies looking for reversible and irreversible effects.

The first two are normally considered together and comprise the dose, whilst the latter is the response. If the period of exposure is fixed and we look at the effect of dose on response then the shape of the resultant curve can tell us much about the way a compound behaves. This is most important when attempting to extrapolate from experimental data to a problem in the field. Some examples of simple dose response curves are shown in Fig. 4.2.

If the dose were to be held constant and the effect of time on response studied, then incremental accumulation of a toxic substance or injury may be seen (Fig. 4.3). The steady accumulation of a toxic compound with no excretion or metabolism by the body can be most closely related to an acute dose assuming accumulation takes place only at the target site. More realistically some metabolism and excretion will occur and a more gradual accumulation result. However, to exert an effect a toxic substance must be able to reach the target site. In some circumstances the toxicant may be accumulated away from the target, such as organochlorine pesticides in fatty tissue, or lead in bone. Until the toxicant is

Ch. 4] The Role of Toxicology

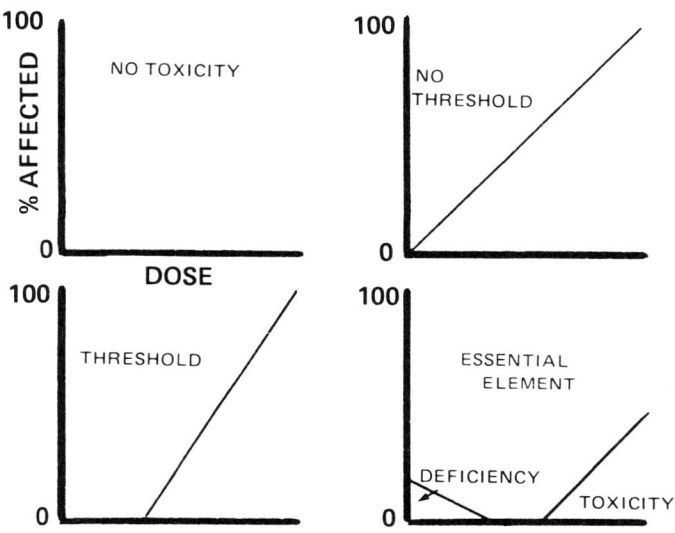

Fig. 4.1 — Dose–effect relationship.

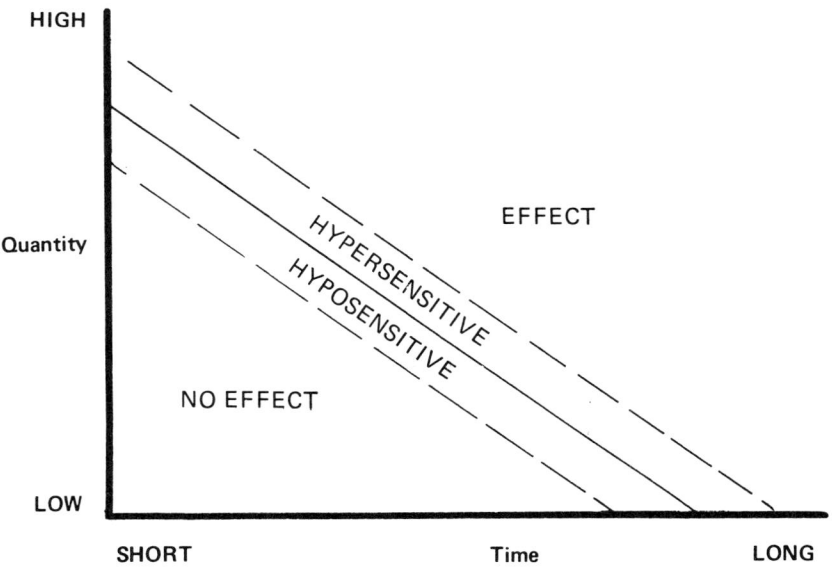

Fig. 4.2 — Some simple dose–response curves.

released and reaches the target, it will not damage the organism. However, if sufficient toxicant can reach the target to cause damage and that damage cannot be completely repaired between successive doses then the residual injury will be accumulated until an effect can be seen.

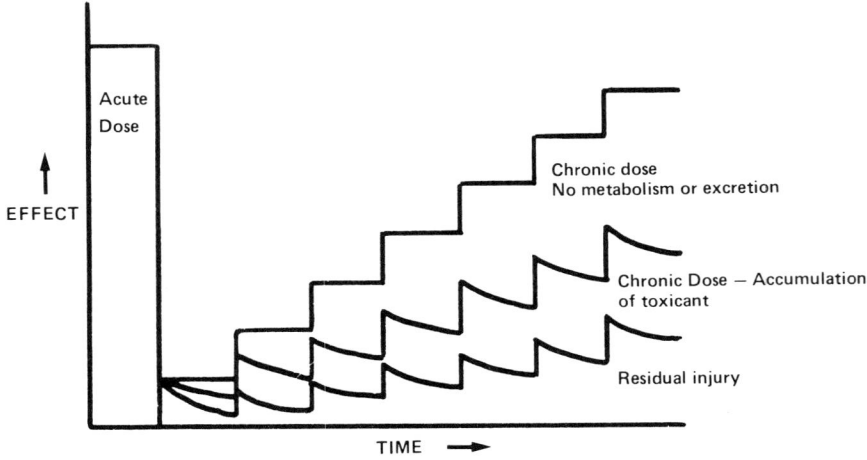

Fig. 4.3 — Accumulation of compound and injury with time (after Klaasen and Doull, 1980 [16]).

As has been mentioned above the problems encountered in the water industry range from those which involve acute (short-term) exposure such as in spills in flowing water to chronic (long-term) exposure such as pipe lining materials. It must be emphasised that acute or short-term data is a poor indicator of chronic toxicity and the data, appropriate to the type of exposure, should be used. For example, the oral LD50 of vinyl chloride in rats is about 500 mg/kg body weight but in a long-term study a dose of 50 mg/kg induced tumours in a high proportion of the animals. The former dose is, in no way, predictive of the latter.

The majority of studies which provide the information on which we make decisions will be those using laboratory animals. However, *in vitro* tests, using bacteria or mammalian cells in culture, are used for assessing mutagenicity. These tests are also employed as short-term screening tests for potential carcinogenicity since the correlation between the ability to induce mutations and carcinogenicity is high.

Sometimes data on the direct effects of a compound on man are available as a result of pharmaceutical and medical research or from industrial toxicology, epidemiology or poisonings, but this information is usually limited.

4.3 HOW GOOD IS THE INFORMATION?

It is important that any evaluation of the toxicology of a compound is not just a data gathering exercise. The experimental work from which the data are derived must also be examined critically. Just as in other branches of science there are good and bad studies and as knowledge increases then studies tend to become more comprehensive and earlier studies viewed more critically. In addition, relatively few studies have been carried out directly in support of environmental toxicology. All experimental work has limitations which does not necessarily invalidate the work but does imply that conclusions drawn from the data must be qualified. The conclusions or interpretation should not exceed the power of the experiment.

Frequently, when assessing toxicological data, we are looking for a no-effect level. It is very important to remember, however, that the no-effect level depends on just how closely one looks for an effect. For example, if at the end of a toxicity study the tissues are examined for macroscopic signs of change but no histopathology is carried out then the no-effect level determined is of very limited reliability compared with a similar study which included histopathology and determined microscopic changes in the tissues. In two life-span studies in rats used to quantify this phenomenon only 70% and 76% respectively of the tumours diagnosed microscopically were observed macroscopically [3]. In other words, about 20% of the data were missing or wrong just in terms of tumour incidence. It is interesting to note that toxicologists are now tending to call the no-effect level the no-observed-effect level.

Carcinogenicity studies are particularly prone to problems, a number of features being relevant.

- The dose may be above the maximum tolerated dose (MTD) and tissue damage may ensue which, in turn, will predispose the animals to tumour formation, particularly in the liver.
- The dose may be so high as to swamp the metabolic system and so overload normal detoxification processes.
- The lifespan of the animals may be shortened, so reducing the period in which tumours can form and increasing the chance of false negative results.
- Lifespan differences may occur between groups making comparisons more difficult. This can be overcome by appropriate statistical analysis.
- Short tests may be carried out which do not take account of the fact that tumours tend to occur later in life and that cancer has a very long induction period.
- Low doses may be used which may give a false negative in a relatively small test population.
- Small groups may be used which decrease the chance of detecting a real effect. With groups of 100 animals and no tumours in the control

group then a tumour incidence of 5% in the test animals is necessary for a significance level of 95%. As the group size decreases so the minimum incidence of tumours necessary to detect a significant difference increases. In the event of the tumour under consideration also occurring in the control groups as part of the natural background then it becomes much more difficult to detect an effect.

There are of course other factors which can influence the results of carcinogenicity studies, such as the purity of the compound used, for example, in the National Cancer Institute trichloroethylene study, the trichloroethylene was stabilised with epichlorohydrin and the latter may have contributed to the positive result [4]. Different vehicles for the test compound may influence the tumour yield as was demonstrated by studies of chloroform in mice in which the test group receiving chloroform in arachis oil showed a significantly greater number of tumours over the test group receiving chloroform in a toothpaste base [5]. Finally, food intake and body weight have been shown to affect the number of tumours so that the incidence of tumours in the controls or body weight differences between groups may make interpretation difficult [6].

There are many factors which can influence toxicity or the expression of toxicity in animals and man (Table 4.2). Arguably the most important of these is pharmacokinetics or toxicokinetics in which the absorption, metabolism and excretion of compounds are taken into account. The compound must enter the body and be transported to the target organ or tissue in the correct form and quantity to cause damage. However, metabolism of the compound can occur by the appropriate cellular enzymes of which the liver is a particularly rich source. Metabolism usually takes place through a chain of enzymes and may act in one of two ways. Firstly, the compound may be detoxified to water-soluble products which can be easily excreted. Alternatively, the compound may be changed to a reactive species and then that compound can express its toxicity. Many of the factors which influence toxicity do so by affecting the absorption, metabolism and excretion of toxicants.

When extrapolating from experimental animals to man the confounding effects of these factors must be carefully considered.

Before assessing critically the toxicological hazard posed by a substance it is necessary to have some information to assess. Unfortunately, for the majority of the vast number of chemicals which exist there is little or no toxicity data. In 1975 a World Health Organisation (WHO) working group on reuse of wastewater [7] noted that of about 300 organic compounds identified in water, acute toxicity data was available for only 45%. Chronic toxicity data was available for an even smaller percentage and carcinogenicity data yet more sparse. At present, with the number of compounds identified in water increasing rapidly, the situation is no better and arguably considerably worse.

Two of the reasons for this state of affairs are cost and time. Toxicology is

Table 4.2
Factors that influence toxicity

- Absorption, Metabolism, Excretion
- Tolerance
- Reserve functional capacity
- Genetic, including Strain or species
- Sex
- Nutrition
- Disease
- Age
- Special physiology (e.g. Pregnancy)
- Life style

expensive particularly in relation to long-term studies. Large numbers of animals need to be maintained and even the more sophisticated procedures tend to be labour intensive using skilled scientists and technicians. In addition, toxicity studies take time, a two-year study takes more than two years to plan and complete. Short-term screening tests are, unfortunately, just that. They may help in the assessment process but are not the whole solution.

There are now a number of good computerised data bases which make access to the published literature much easier than ever before. However, it must be emphasised that a major portion of toxicity data is not published but is held by industrial organisations in confidence because of its commercial significance. This material is not generally or readily available.

4.4 THE USE OF TOXICITY DATA IN ENVIRONMENTAL PROTECTION

Toxicity is the intrinsic capacity of a compound to cause injury. Hazard is the capacity of that compound to cause injury under the circumstances of exposure. The presence of a toxic substance does not *per se* mean that a hazard exists.

When using toxicology in environmental protection, particularly when dealing with water, the assessment of exposure is very important. It must be remembered that exposure from other sources such as food and air may be as significant or even substantially greater. For example, the estimated daily intake of benzene is about 240 micrograms of which less than 1 microgram will normally come from water [8]. A similar proportional difference exists for PAH [9].

Unfortunately man is not a big rat. Therefore the process of extrapolating from the data available on the toxicity of a compound in laboratory animals requires care. Often judgements must be made before all the information and understanding we need are available. There are ways in which this can be achieved successfully so long as the constraints and limitations on the results are borne in mind. However, the toxicologist must be prepared to enter the realm of 'practical politics'.

The most commonly used technique up to a few years ago was to apply 'safety factors'. A preferable term is 'uncertainty factors' since they are arbitrary factors selected to reflect uncertainty in the data and the way it relates to man. Previously, organisations such as the United Nations Food and Agriculture Organisation (FAO) used a factor of 100 for their Acceptability Daily Intakes (ADI) for food contaminants. This really consisted of two factors, 10 to reflect a possible increased sensitivity of man relative to laboratory animals, and an additional 10 to account for the potential range of sensitivity to be found in a heterogeneous human population. If insufficient data were available then no ADI was set. Unfortunately, decisions may have to be taken no matter how little the available data is or how large an uncertainty factor may need to be employed. There can, however, be logic in the application of these factors as shown in the methods used by the US National Academy of Sciences (Table 4.3) [10]. For example, suggested no adverse response levels have been calculated for toluene to indicate safe levels if exposure of 24 hours and 7 days were to occur via drinking water following a spill. Sufficient acute data was available from animals to justify a factor of 100 and a figure of 420 mg/l for 24 hours was suggested. The limited sub-acute data warranted a factor of 1000 and a figure of 35 mg/l for 7 day exposure resulted [10].

Table 4.3
'Safety' or 'uncertainty' factors (after NAS (1980))

	Factor
Good chronic or acute human data + chronic or acute data in other species	10
Good chronic or acute toxicity data one or more species	100
Limited or incomplete acute or chronic data	1000

More recently there has been a number of attempts to produce mathematical models to establish a safe level for exposure to carcinogens. Such models are designed to extrapolate from the results of high dose experiments in laboratory animals to environmental levels of the chemical under review. Unfortunately, it is early days for these models and many uncertainties remain. The use of these models has been summarised by Hoel who said:

> recently, interest has been directed to statistical models of dose response for estimating low dose risk based upon high dose data. Curve fitting has been

commonplace using functions such as probit, logit, multi-hit, multi-stage, etc. Some of these functions have a tradition in bio-assay work, while others attempt crudely to describe biological mechanisms. For example, the multi-stage model assumes the carcinogenesis process is represented by a direct acting carcinogen interacting with DNA as in a single cell somatic mutation theory. This simple model does not include consideration of DNA repair, the immune surveillance system, genetic or environmental susceptibilities in the population, or pharmacokinetics [11].

Considerable improvements are likely to be made in these models in the not too distant future as a result of the increased interest in the field. In addition, the availability of the data from recent, extremely large carcinogenicity studies [12] which were able to use a greater range of doses going to lower levels than hitherto possible, will help to reduce the uncertainty about the shape of the dose response curve at these lower levels.

In the real world decisions need to be made and imperfect information is better than no information at all. These limitations need to be recognised in making such decisions. Because of the complexities of toxicology and the potential consequences of a decision which errs on the wrong side of safety, there is sometimes a tendency to clutch at standards for environmental pollutants as a permanent way among shifting sands. Standards, however, do not constitute a hard and fast line below which there is no effect and above which effects will be seen to gradually increase over a range of concentrations actually starting well above the standard. Numbers in toxicology are not constants, they will shift with the changing circumstances of their derivation. Consequently standards will alter in response to both our increasing knowledge and the changing circumstances of exposure. The efforts of WHO to introduce guidelines for water contaminants which are not hard and fast figures does constitute a significant and useful development in this complex area.

A good example of their approach are trichloroethylene and tetrachloroethylene for which there was a clear need for guidance regarding what would be considered a safe level in groundwater. WHO have recognised the difficulties and uncertainties in extrapolating from the lowest dose in tumour studies of 869 mg/kg/day for trichloroethylene and 368 mg/kg/day for tetrachloroethylene to environmental concentrations of about 100 and 10μg/l respectively [13, 14].

These guidelines also show a change in philosophy from many previous standards such as the standard for lead for which the weight of evidence for toxic effects from environmental levels of exposure is much more substantial even though there are still uncertainties.

4.5 RESEARCH NEEDS

In all this uncertainty the problems of mixtures have not been mentioned. Almost invariably in environmental toxicology we are dealing with exposure to more than one compound at a time. When we examine organic micropollutants in water, for example, then the mixture is extremely complex [15]. It is highly probably that additive, synergistic and antagonistic effects do occur in this brew of chemicals all present in very small quantities but our knowledge is at best extremely limited and at worst virtually non-existent. In addition, as has been mentioned previously, our knowledge of the behaviour of very low doses of compounds in the intact animal is also limited. Included in this is our ignorance on effects which can result from long-term exposure to these low concentrations. There are many different biochemical processes, even in the single cell, which are dependent on each other, and the efficient operation of each process is important. Consequently, effects which cannot be readily identified but which result from a long term disturbance of some of these systems could be very important in their effects on disease patterns and susceptibility to disease in the population as a whole. Changes in enzyme activity in the cells can lead to their less efficient functioning as can depletion of those molecules which protect the cell against reactive foreign compounds. Low level effects on the endocrine organs, which produce the chemical regulators of body functions, either directly, by stimulating hormone secretion, or indirectly, by stimulating hormone section, or indirectly, by affecting the biochemical systems responsible for endocrine control, may produce long-term, low level disturbance of one of the major biological control mechanisms of the body. It must be appreciated that any effects on health which do occur as a result of these metabolic changes are more likely to be superimposed on the 'normal' pattern of disease rather than be novel and, as such, will be all the more difficult to discern.

Clearly there is a need for fundamental research in these areas with particular emphasis being given to mechanisms of toxicity. Such studies will help us predict, with much more accuracy, what will happen in a given situation. Unfortunately, the greatest proportion by far of the money spent on toxicology is in carrying out routine tests to satisfy regulatory authorities rather than understanding the how, why and what of toxicity.

In view of the numerous compounds which appear in the environment of which we have extremely little, if any, knowledge about toxicity, two additional areas of research are very important. One is the development of short-term *in vitro* screening tests. However, the use of these tests must go hand in hand with fundamental mechanistic studies *in vivo* or the only result will be a confusing mass of uninterpretable data. The second is the development of more sophisticated techniques of measuring structure activity relationships in order to develop models for the prediction of toxicity from chemical structure.

Improvements in mathematical models for extrapolating from high to low doses and from species to species are most important for environmental toxicology and not just for carcinogens. Such developments are essential if we are to help in solving real problems rather than simply providing legislators with more accurate figures.

4.6 DISCUSSION

Toxicology is the applied science which brings together many disciplines in studying the adverse effects of chemicals on living organisms. This subject has an important role in environmental protection in helping to determine the acceptability of materials and chemicals to be used and in helping to assess the extent of action required by the presence of pollutants. The approach can vary from 'will there be a problem if we do this' to 'we have found this, is there a problem and how big?'.

There is a range of basic information regarding a compound which is needed to make a comprehensive evaluation of toxicity but this information is often lacking or difficult to obtain. In assessing whether a hazard exists, it is necessary to demonstrate exposure and the probability of toxicity at those levels of exposure. At present methods of extrapolating from high to low doses and from species to species have many limitations. There is an urgent requirement for fundamental research which will enable more accurate predictions of the effects of low levels of exposure to individual compounds and to complex mixtures.

The requirement for toxicologists to contribute to decisions which must be made in environmental protection has been the impetus for a practical and common sense approach and the stimulus for much needed research.

4.7 REFERENCES

[1] Sharratt, M. Uncertainties associated with the evaluation of the health hazards of environmental chemicals from toxicological data. In *The Evaluation of Toxicological Data for the Protection of Public Health.* Proceedings of the International Colloquium Luxembourg, 1976, Pergamon Press.

[2] Paracelsus (Theophrastus ex Hokenheim Cremita): Von der Besucht. Dillingen 1567.

[3] Kulwich, B. A., Hardistry, J. F., Gilmore, C. E. and Ward, J. M. (1980) Correlation between gross observations of tumours and neoplasms diagnosed microscopically in carcinogenesis bio-assays in rats. *Journal of Environmental Pathology and Toxicology*, 3, 281–287.

[4] National Cancer Institute. (1976) Carcinogenesis bio-assay of trichloroethylene Cas/no. 79-01-6 NCI-CG-TR 2. Washington DC.

[5] Rowe, F. J. C., Palmer, A. K., Worden, A. N., and Van Abbe, N. J. (1979) Safety evaluation of toothpaste containing chloroform 1. Long term studies in mice. *Journal of Environmental Pathology and Toxicology.* 2, 799–819.

[6] Tucker, M. J. (1979) The effect of long term food restriction on tumours in rodents. *International Journal of Cancer.* 23(6), 803–807.

[7] World Health Organisation. (1975) International Reference Centre for Community Water Supply. Technical Paper No. 7. *Health effects relating to direct and indirect re-use of waste water for human consumption*, WHO.

[8] Health and Welfare Canada. (1979) Benzene. Information Directorate, Department of National Health and Welfare, Ottawa. 79-EHD-40.

[9] Shabad, L. M. and Il'nitshii, A. P. (1980) Gigiena sanitaria Vol. 35 p. 268, 1970. Cited in *Polycyclic Aromatic Hydrocarbons*. Health and Welfare Canada, Ottawa. 80-EDH-50.

[10] United States National Academy of Sciences. (1980) *Drinking Water and Health*, Vol. 3. National Academy Press. Washington DC.

[11] Hoel, D. G. (1979) Statistical approaches to toxicological data. *Environmental Health Perspectives*, 32, 267–271.

[12] Gaylor, D. W. (1979) The ED_{01} Study: Summary and Conclusions. *Journal of Environmental Pathology and Toxicology*. 3, 179–183.

[13] World Health Organization. *Guidelines for Drinking Water Quality* (WHO, Geneva, in press).

[14] Weisburger, E. K. (1977) Carcinogenicity studies on halogenated hydrocarbons. *Environmental Health Perspectives*. 21, 7–16.

[15] Fawell, J. K. and James, H. A. (1981) Problems of assessing the toxicological significance of organic micro-pollutants in drinking water. *Proceedings of the Institute of Biology Symposium Organic Micro-pollutants in Water.* March 13, London.

[16] Klaasen, C. D. and Doull, J. (1980) Evaluation of safety: toxicologic evaluation. In: *Toxicology (The Basic Science of Poisons)* 2nd edn, Macmillan.

CHAPTER 5

Epidemiology and water quality

PATRICIA FRASER
London School of Hygiene and Tropical Medicine

5.1 WILLIAM FARR AND CHOLERA

William Farr (1807 to 1883) died one hundred years ago. He was the founder of medical statistics as we know it, and during his 40 years at the General Register Office he developed and analysed new sources of statistical information in the fields of public health and social welfare. Farr's 'Letters to the Registrar General', together with his many other publications, give a clear impression of the major health problems at that time. Thus it is appropriate in this year commemorating a life and work of William Farr, that I should begin my examination of the ability of an epidemiological approach to some answer some specific questions about water quality, with a reference to one of the better-known examples of his work, recently reviewed by Lewes [1] — the study of cholera in London in the mid-nineteenth century.

There were two main theories of its transmission. The first was that it was largely caused by airborne poisons, and the second was that cholera was contagious, the infection being passed when contaminated human excreta were discharged into rivers and subsequently pumped into water mains, or found their way into wells or other sources of drinking water. This latter theory, particularly associated with John Snow, focused policies for preventing the spread of the disease on water supply.

Farr was aware that through its system of death registration and the publication of Weekly Returns, the General Register Office held a reliable source of information, which in London could be made available in time to be of use in ascertaining the spread of an epidemic and the effectiveness of measures for its control. He provided Snow with figures of mortality from cholera in London Registration Districts, listed according to the water company responsible for supply. They showed that 'mortality from cholera was lowest in the districts which have their waters from the Thames so high in its course as Hammersmith or Kew ... and greatest in the districts which derive their water from the Thames so low as Battersea and Hungerford Bridge'.

Whereas in most parts of London individual water companies had monopoly powers of supply, over a large area of South London water was supplied in competition by two companies. The actual houses supplied by each company were almost randomly distributed throughout all the streets of the area and hence different only in their water supply. Using death notifications provided by Farr, Snow discovered as many addresses as possible where cholera deaths had occurred and where it was known which company supplied the water. From this information, he estimated that there had been 57 deaths per 1000 houses supplied by the Southwark and Vauxhall Company drawing its water from the polluted tidal waters of the Thames, against only 11 per 1000 in those supplied by the Lambeth Company using an unpolluted supply from above the level of the tides. Snow's final report suggests that the contrast was even greater, the ratio of deaths from cholera per 1000 houses supplied by the two companies being 14:1.

This study provided strong evidence in support of the theory of waterborne transmission, although it was not until the advent of more powerful microscopes in the 1890s that the causative organism was isolated. The importance of Farr's and Snow's work lies in the fact that it offered a practical solution to the spread of cholera. Their work demonstrated too that a death registration system could be used both as an early warning for the spread of disease and as a powerful analytical tool.

5.2 SWIMMING-ASSOCIATED ILLNESS

In the century which has elapsed since William Farr's death, advances in knowledge of effects of constituents of water on human health have been made, and in moving in the developed countries from an era of waterborne diseases to one of reliance on wholesome water supplies which can be consumed safely, there has been remarkable progress [2]. However, while in the developed world potable water of an acceptable and reliable quality is now demanded as a right, several investigators have suggested that there may be measurable health effects associated with swimming in sewage-polluted waters. Recently, prospective cohort studies carried out in the US have explored the relationship between the extent of pollution of bathing beaches, as measured by mean enterococcus density, and the incidence of mild gastro-intestinal disease in the swimming population [3]. The studies were motivated by the desire for objective standards for marine recreational bathing, based on the concept of acceptable risk. That is to say, if the safety of bathing is the main consideration, then the guidelines should be framed in such a way that people who bathe in waters which comply with them take only a known risk of incurring one or more of a specified set of diseases, the guide levels being chosen so that the risk is less than a predetermined probability [4].

The Cabelli studies were undertaken at three locations in the US differing in temperature, salinity, tides and sources of pollution. The study design comprised a series of 'trials', a trial being a visit to a particular beach on a particular day, to obtain measurements of enterococcus density and to recruit both swimmers and non-swimmers to the study. Those who agreed to participate were telephoned subsequently to find out whether they had experienced any gastro-intestinal symptoms in the following week.

Cabelli *et al.* felt that their results demonstrated a direct linear relationship between swimming-associated gastro-intestinal illness and the quality of the bathing water. However, the subjective nature of disease reporting and failure to seek confirmation of the diagnosis through the isolation of any causative organisms, weakens the confidence which can be placed in this conclusion. While a linear trend was demonstrable on pooling the data from all locations, the limited data available from individual sites do not support the hypothesis that the same water quality indicator-disease relationship is applicable for all beaches and bathing populations [4]. The Cabelli studies were large in their scale of data collection, but provided only a modest amount of information. Thus while the concept of acceptable risk is commendable, the magnitude of the prospective studies which would be required to generate a firmer basis for prediction is formidable.

D'Alessio *et al.* used instead a retrospective case-control approach as a first step in their investigation of the hypothesis that transmission of enteroviruses via recreational water might make a regular contribution to the usual enterovirus disease patterns, yet be indiscernible because discrete, recognisable outbreaks do not occur [5]. They compared 296 children with enteroviral-like syndromes with 679 well children, with respect to the frequency and location of swimming in the preceding two weeks. Specimens for virus culture were taken from 241 ill patients and 27 controls.

With an odds ratio of 3.4, exclusive beach swimmers were found to have a significantly increased risk of enteroviral illness, whereas bathing exclusively in chlorinated swimming pools carried no increased risk. Likely sources of bias would tend to make this estimate conservative, thus D'Alessio *et al.* concluded that they had demonstrated a significantly increased risk of enterovirus illness in children swimming at unpolluted beaches which warrants further study.

5.3 WATER HARDNESS AND CARDIOVASCULAR DISEASE

While the main consideration in ensuring the safety of public water supplies is still the removal of bacterial contamination, the health aspects of chemicals in drinking water have received attention in recent years. For example, over the last 20 years there has been considerable research into the relation between hardness of drinking water and cardiovascular disease. Several statistical studies have demonstrated a highly significant inverse association although the precise nature of the water components involved is still not known. The most recent of these investigations, Phase 1 of the British Regional Heart Study, was undertaken to try to explain the substantial regional variations in coronary artery disease and stroke in Great Britain [6,7]. This study was more extensive and used more reliable data on water quality than previous British studies, and hence the authors were able to examine the form of the association between water hardness and cardiovascular mortality in greater detail. After testing many different multiple regression models, they concluded that there were five variables that collectively had a highly significant effect on cardiovascular mortality as measured by the standardised mortality ratio. These were water hardness, percentage of days with rain, mean daily maximum temperature, percentage of manual workers, and car ownership. This five-variable model accounted for 78 per cent of the regional variation in cardiovascular mortality, each variable making a separate and important contribution which could not be attributed to its association with other variables in the study.

Further investigation showed that the effect of water hardness was non-linear, being much greater in the range from very soft to medium-hard than from medium to very hard water. Thus after adjusting for climatic and socioeconomic factors, on average very soft water towns were found to have about ten per cent higher cardiovascular mortality than medium-hard or harder water towns.

In this retrospective geographic study, no single water factor could be blamed for the increase in cardiovascular mortality since many of the water parameters were highly correlated with one another. At this stage in the British Regional Heart Study it can only be said that cardiovascular disease is influenced by water hardness, or by some factor closely associated with it, which could be either a harmful factor in soft water or a protective factor in hard water. The effect of the water factor is small, compared with other cardiovascular risk factors and there does not appear to be any justification at present for recommending water hardening [7]. Phase 2, a cross-sectional clinical survey in middle-aged men in 24 British towns, and Phase 3, a prospective study of the incidence of cardiovascular disease in these men, are expected to provide firmer conclusions regarding the causal role of the many individual and environmental risk factors being studied.

5.4 NITRATES AND INFANTILE METHAEMOGLOBINAEMIA

The occurrence of methaemoglobinaemia in young bottle-fed infants whose feeds have been made-up with nitrate-rich water is one of the few examples where there is definite evidence that consumption of water has led to ill-health [8]. The condition has been recognised since 1945 and clinical cases are rare in countries with a high standard of public health. Of the 2000 cases reported in the world literature since 1945, most have been caused by drinking water from private and often bacterially contaminated wells rather than from public water supplies.

The World Health Organization (WHO) standards for nitrate in drinking water [9] were established largely to guard against the development of infantile methaemoglobinaemia. They are based on an analysis of reported cases in relation to the level of nitrate in the water associated with each case. Surveys suggested that the majority of cases of methaemoglobinaemia have occurred when nitrate levels have exceeded 100 mg/l and that cases are rare when nitrate levels are less then 45 to 50 mg/l [8]. However, in these surveys the nitrate concentration was often not known, and even the available data on nitrate levels could be misleading when the water samples for analysis were obtained some time after the acute illness, during which the nitrate concentration may have changed considerably. Information about possible contributory factors may also be lacking.

Thus while the object of the surveys was to define the level of detectable risk, deficiencies in the data available leave doubts about the correlation between the concentration of waterbourne nitrate and the frequency of cases of infantile methaemoglobinaemia. In reviewing the situation in 1974 [8], the International Standing Committee on Water Quality and Treatment concluded that there was insufficient evidence to permit raising the recommended level above 45 to 50 mg/l, but that maintenance of this standard would provide a margin of safety against the many aggravating factors which might interact in the development of a case of infantile methaemoglobinaemia.

5.5 NITRATES AND STOMACH CANCER

The fact that the standards for nitrate in drinking water were established to guard against the development of methaemoglobinaemia and not with any hypothetical cancer risk in mind, has been overlooked in much of the rather alarmist reporting of late which has tended to equate nitrate levels over 50 mg/l with an increased risk of stomach cancer. Drinking water nitrate levels are rising in parts of Britain and have given cause for concern because under certain circumstances nitrates, which are derived mainly from bacterial reduction of ingested nitrates, may react *in vivo* with nitrosatable substrates in certain foods to form N-nitroso compounds. Many of these substances have been found to be carcinogenic when administered to laboratory animals, but as yet they have not been definitely incriminated as the cause of any human cancer [10].

Reviews of epidemiological studies prior to 1979 concluded that there was no evidence at the time unambiguously associating nitrates, nitrites, or N-nitroso compounds with cancer of any organ in man [11, 12]. Since then further studies of the relationship between nitrate levels in drinking water and stomach cancer have been completed in England [13, 14, 15, 16], Chile [17], Hungary [18],

Italy [19], Denmark [20], and France [21]. None of these recent studies provides convincing evidence of a link between waterborne nitrate and stomach cancer [22] which is in fact declining worldwide. The studies by Davies and Beresford in England, by Zaldivar and Wetterstrand in Chile, and by Vincent *et al.* in France are negative; our findings in eastern England and those of Juhasz *et al.* in Hungary are inconsistent, and no inference can be drawn from the Italian study in the absence of a low-risk group for comparison. While Jensen suggested tentatively that his results in two Danish towns support a possible weak role for nitrate in the aetiology of stomach cancer, lack of definite evidence of higher nitrate intake in the town with a higher incidence of stomach cancer weakens this conclusion. In fact, lack of information on total nitrate intake in populations at differing risk, or individuals with and without stomach cancer, is a weakness of many epidemiological studies in this area.

While concern is often expressed over rising levels of nitrate in drinking water, consideration is seldom given to the contribution made by water to total nitrate intake. A recent population survey of dietary nitrate in well-water users in Norfolk has provided some information on the relative importance of water and food as sources of nitrate [23]. The well-water nitrate levels ranged from 0 to 269 mg/l many wells having substantially higher levels than the public water supplies in the same area. Using 24-hour urinary nitrate excretion as a surrogate measure of total nitrate intake, and an assay of the nitrate content of drinking water in conjunction with a diary record of consumption, it was possible to obtain an estimate of nitrate intake from water. The diet diary also provided a rough estimate of nitrate intake from solid food. From these data the contributions of waterborne nitrate and food to total nitrate intake could be assessed over a wide range of concentrations of nitrate in drinking water.

Where the waterborne nitrate level was less than 50 mg/l 30 per cent of ingested nitrate was from water. As the well-water nitrate concentration rose, the contribution of water to daily intake incdeased; at levels between 50 and 100 mg/l, on average, nearly 70 per cent of daily intake was from water, and above 100 mg/l over 80 per cent of daily intake was waterborne. Thus it is only at levels well above those currently recommended that waterborne nitrate is the major contributor to total nitrate intake.

In the recent studies examining the relationship between stomach cancer and nitrate in drinking water, with the exception of the Hungarian study, nitrate levels rarely exceeded 50 mg/l. Thus, based on our finding in Norfolk, the contribution of water to daily nitrate intake would be no more than one third in persons consuming a normal Western diet, and food would be the more important source of nitrate in the populations studied.

5.6 FLUORIDATION AND CANCER

The longest and costliest case in Scottish legal history was concluded in July 1983 with a complete vindication of the safety and efficacy of fluoridation. The controversy over the safety of fluoridation arose when Yiamouyiannis and Burk purported to show that cancer mortality in US fluoridated communities was greater than in communities without fluoridated water supplies [24]. This finding has not been confirmed in more detailed analyses of a wealth of vital statistical data by several investigators. For example, a comparison of time trends in site-specific standardised cancer mortality ratios in the fluoridated and non-fluoridated cities provided no evidence that fluoridation of water supplies was associated with an increased risk of cancer at any of the sites examined [25]. There is evidence, however, that the addition of fluoride to drinking water so that the concentration is raised to 1 ppm has a beneficial effect on the incidence of dental caries in children [26]. As in Farr's day, an epidemiological approach and the careful analysis of appropriate statistics has proved invaluable in resolving an important issue regarding the safety of public water supplies.

• The author is supported by the Medical Research Council.

5.7 REFERENCES

[1] Lewes, F. (1983) William Farr and cholera. *Population Trends.* **31**, 8–12.
[2] Goodman, A. H. Potable water quality. In: *Developments in Water Treatment – 1.* Edited by W. M. Lewis. Chap. 1 pp. 1–32. Applied Science Publishers Ltd, England, 1980.
[3] Cabelli, V. J., Dufour, A. P., McCabe, L. J., and Levin, M. A. (1982) Swimming-associated gastro-enteritis and water quality. *Am. J. Epidemiol*, **115**, 606–616.
[4] Lacey, R. F. (1982) Comments on the Cabelli studies. Report prepared for the Department of the Environment Standing Technical Advisory Committee on Water Quality, Water Research Centre Environmental Protection.
[5] D'Alessio, D. J., Minor, T. E., Allen, C. I., Tsiatis, A. A., and Nelson, D. B. (1981) A study of the proportions of swimmers among well controls and children with enterovirus-like illness shedding or not shedding an enterovirus. *Am. J. Epidemiol*, **113**, 533–541.
[6] Pocock, S. J., Shaper, A. G., Cook, D. G., Packham, R. F., Lacey, R. F., Powell, P., and Russell, P. F. (1980) British Regional Heart Study: geographic variations in cardiovascular mortality, and the role of water quality *Brit. med. J.*, **280**, 1243–1249.
[7] Shaper, A. G., Pocock, S. J., Packham, R. F., Lacey, R. F., and Powell, P. (1983) Softness of drinking water and cardiovascular disease – practical implications of recent research. *Health Trends*, 15, 22–24.
[8] Report by the International Standing Committee on Water Quality and Treatment. Nitrates in Water Supplies. *Aqua*, 1974, **1**, 5–24.

[9] World Health Organization. (1971) International Standards for Drinking Water. 3rd edn. World Health Organization, Geneva.
[10] National Academy of Sciences. (1981) *The Health Effects of Nitrate, Nitrite and N-Nitroso Compounds*. National Academy Press, Washington DC.
[11] Royal Commission on Environmental Pollution. (1979) 7th Report, Agriculture and Pollution, Chap. IV, pp. 87–125, HMSO, London.
[12] Fraser, P., Chilvers, C., Beral, V., and Hill, M. J. (1980) Nitrate and human cancer: a review of the evidence. *Int. J. Epidemiol.*, **9**, 3–11.
[13] Fraser, P., and Chilvers, C. (1981) Health aspects of nitrate in drinking water. *The Science of the Total Environment*, **18**, 103–116.
[14] Davies, J. M. (1980) Stomach cancer mortality in Worksop and other Nottinghamshire mining towns. *Br. J. Cancer*, **41**, 438–445.
[15] Beresford, S. A. A. (1981) The relationship between water quality and health in the London area. *Int. J. Epidemiol.*, **10**, 103–115.
[16] Beresford S. A. A. Is nitrate in the drinking water associated with the risk of stomach cancer in the urban UK? *Int. J. Epidemiol.* (in press).
[17] Zaldivar, R., and Wetterstrand, W. H. (1978) Nitrate nitrogen levels in drinking water of urban areas with high- and low-risk populations for stomach cancer: an environment epidemiology study. *Z. Krebsforsch*, **92**, 227–234.
[18] Juhasz, L., Hill, M. J., and Nagy. G. (1980) *Possible Relationship Between Nitrate in Drinking Water and Incidence of Stomach Cancer.* International Agency for Research on Cancer Scientific Publication No. 31, pp. 619–623, IARC, Lyon.
[19] Amadori, D., Ravaioli, A., Gardini, A., Liverani, M., Zoli, W., Tonelli, B., Ridolfi, R., and Gentilini, P. (1980) N-nitroso compound precursors and gastric cancer: preliminary data of a study on a group of farm workers. *Tumori*, **66**, 145–152.
[20] Jensen, O. M., (1982) Nitrate in drinking water and cancer in Northern Jutland, Denmark, with special reference to stomach cancer. *Ecotoxicology and Environmental Safety.* **6**, 258–267.
[21] Vincent, P., Dubois, G., and Leclerc, H. (1983) Nitrate in drinking water and cancer mortality. Epidemiological study in the north of France. *Rev. Epidém et Santé Publique.* **31**, 199–207.
[22] Fraser, P. (1983) Nitrates: Epidemiological Evidence. In *Symposium on Interpretation of Epidemiological Evidence with Special Reference to Evidence Suggesting Lack of Carcinogenicity,* Oxford, IARC Sci. Publ. (in press).
[23] Chilvers, C., Inskip, H., Caygill, C., Bartholomew, B., Fraser, P., and Hill, M. A survey of dietary nitrate in well-water users (submitted for publication).
[24] Yiamouyiannis, J., and Burk, D., (1977) Fluoridation and cancer. Age-dependence of cancer mortality related to artificial fluoridation. *Fluoride.* **10**, 102–123.

[25] Chilvers, C. (1982) Cancer mortality by site and fluoridation of water supplies. *J. Epidemiol Community Health.* **36**, 237–243.
[26] World Health Organization. (1969) Fluoride and dental health. *WHO Chronicle.* **23**, 505–512.

CHAPTER 6

Ecotoxicology and environmental quality

W. H. KÖNEMANN
Ministry of Housing, Physical Planning and Environment, The Netherlands

6.1 ECOTOXICOLOGY AND ENVIRONMENTAL QUALITY

After the rapid growth of mammalian toxicology since World War II, a new branch of toxicology has developed mainly in the last 20 years, greatly stimulated by the observation of unwanted side-effects of pesticides and the disappearance of many fish species from our major rivers. Obviously the deterioration of the quality of the environment has stimulated the development of ecotoxicology. The question, however, is to what extent ecotoxicology helps to improve or defend the quality of the environment.

Although some of the observations below may be also valid for other compartments of the environment, I would like to confine myself to the aquatic ecosystem. The expression ecotoxicology as used in this chapter addresses only aquatic toxicology, which is also the most developed part of ecotoxicology. Ecotoxicology can have several goals, such as:

- to predict which levels of chemicals do not have a negative impact on environmental quality;
- to identify chemicals responsible for poor environmental quality;
- to monitor the actual quality of surface water.

Although monitoring the actual quality of surface water is an interesting field of work, it differs in too many aspects from the routinely applied ecotoxicology, to deal with in this chapter. To some extent the same holds for the identification of chemicals responsible for a bad environmental quality. This task is nearly always complicated by uncertainty about the completeness of the chemical analyses of the surface water under investigation. Furthermore it is probably easier for an ecotoxicologist to predict which level of a chemical is safe, than to indicate the degree of damage inflicted by a certain concentration of a chemical to the environment. Therefore I will mainly limit myself to the first goal, the prediction of the levels of chemicals which do not have a negative impact on the environment, or in the ecotoxicological jargon, the establishment of no-effect-levels (NELs).

6.2 CURRENT ECOTOXICITY TESTS

Having reduced my subject so far, I shall try to show the complexity of what remains starting with a short overview of tests. I am relying on the discussion within the OECD Programme for Updating of Test Guidelines as a basis for this overview.

The most generally applied tests are of a short duration:

LC 50 test with fish — usually 4 days.
> Mortality is the most important criterion, but additional observations may help to determine a semi-quantitative no-effect-level (macroscopic effects, not statistically evaluated).

EC 50 test with *Daphnia sp.*: 1 or 2 days.
> Immobilisation is almost the only criterion.

Reproduction toxicity test with *Daphnia magna*: 14–21 days.
> A NEL for reproduction is established.

Algal growth test: 3 or 4 days.
> In spite of its short duration this test can be considered to be chronic, covering all parts of the life cycle of the test species.

All of these tests are applied routinely. The last two tests cause the experimenter some brain-racking now and then, because of problems with the suitability of the medium, but these are not insurmountable. All of these tests are included in the OECD Test Guidelines [1]. Similar types of tests with other organisms (snails etc.) are equally feasible but not routinely applied.

Aquatic tests for which guidelines are proposed within the OECD are:

14 days LC 50 with fish.
> From a scientific point of view this test clearly does not provide much additional information compared with the existing 4 days LC 50 [2].

Early life stage test with fish.
> Exposing fish eggs immediately after fertilisation and following their development until the stage of swimming fry. A no-effect-level is determined. It has been shown that in many cases this development period is the most sensitive stage in the life of a fish [3].

The above-mentioned tests are certainly not the only ecotoxicity tests which have been documented in literature, but these are the ones most frequently applied and in use in many countries.

Now the question arises how and if with these tools an NEL of a chemical for the aquatic environment can be established. As a first approximation I will assume that this is possible and that the lowest NEL of a set of individual tests may serve as the NEL for the environment. To support this assumption I will discuss the relative value of different types of tests.

Slooff et al. [4] have published the results of a very interesting study comparing the susceptibility of various freshwater species to chemicals under laboratory test conditions on the basis of the results of toxicity tests with 22 aquatic species (bacteria, algae, protozoa, crustaceans, insects, coelenterates, molluscs, fishes and amphibians) of different trophic levels. It was shown that the sensitivity of species for one chemical may vary up to a factor of a thousand or more. Slooff, however, also calculated the average susceptibilities based on no-observed-lethal or effect-concentrations (NOL(E)C values). His findings are given in Table 6.1.

Table 6.1
The relative susceptibility of test species based on toxicity data on compounds

Test species		Test species	
Pseudomonas putida	4.0	*Aedes aegypti*	4.6
Microcystis aeruginosa	1.2	*Culex pipiens*	6.3
Chlorella pyrenoidosa	4.2	*Lymnaea stagnalis*	2.9
Scenedesmus pannonicus[†]	4.2	*Hydra oligactis*	4.6
Scenedesmus pannonicus[‡]	2.0	*Leuciscus idus*	3.5[‡]
Selenastrum capricornutum	3.5	*Salmo gairdneri*	1.7
Entosiphon sulcatum	2.0	*Poecilia reticulata*	5.3
Uronema parduczi	5.3	*Oryzias latipes*	5.3
Chilomonas paramecium	3.2	*Pimephales promelas*	1.8
Daphnia magna	1.4	*Xenopus laevis*	3.7
Daphnia pulex	1.0	*Ambystoma mexicanum*	2.7
Daphnia cucullata	1.2[†]		

[†] Derived from LC50 values for daphnids.
[‡] Derived from LC50 values for fishes.

This table already shows what is completely clear from his integral paper: not one species is clearly more or less susceptible to a wide range of chemicals than other species. This is an important result, since it saves us difficult discussions about the choice of test species for ecotoxicity testing.

The great variability of NELs of different test species for one certain chemical points at the need to test more than one species. To help to determine how many species should be tested in order to obtain a reasonable estimate of the NEL for the environment, I have calculated from Slooff's data the ratio between the lowest NEL in a given set of tests (LNEL set) and in total number of 22 tests (LNEL total). In Fig. 6.1 the results of these calculations are presented.

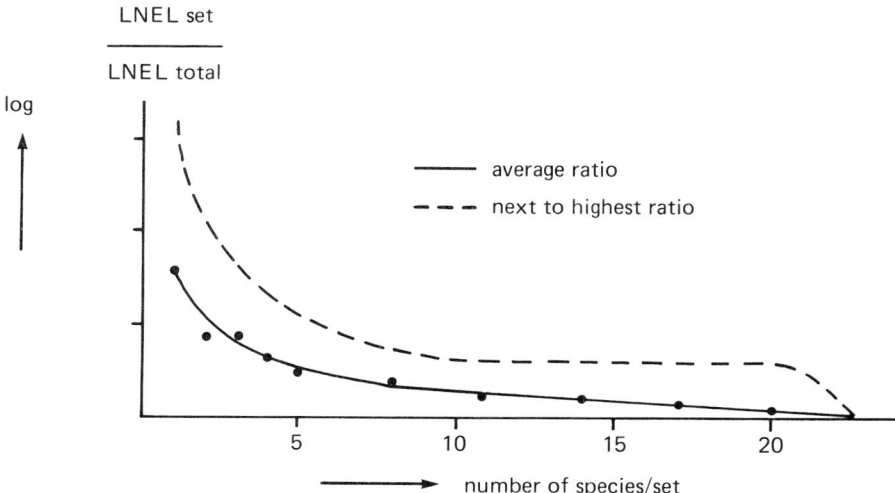

Fig. 6.1 – Decrease of lowest NEL with increasing number of species tested (for explanation see text).

The sets of test species for which the calculations have been performed are based on common practice. The first data point (1 species/set) is based on the toxicity to *Oryzias latipes* (rice fish), then *Daphnia magna* (waterflea) is added, next *Selenastrum capricornutum* (an algal species) and so on. The choice of the tests is somewhat arbitrary, but since the susceptibilities of the test species vary only slightly, it will not affect the outcome. In Fig. 6.1 the drawn line represents the average ratio for all 14 chemicals and the broken line the next to highest ratio for these chemicals.

Assuming for the time being that the 14 chemicals and the 22 test species are a proper representation of all chemicals and species, some conclusions can be drawn:

- It is likely that testing 2 or 3 species gives a good indication of the order of magnitude of the lowest NEL for acute toxicity to a wide range of species.
- Testing of 3 or 4 more species improves the reliability of an estimate of the lowest NEL. There is no sense in testing yet more species, since the reliability of the estimate hardly increases.

The first weakness in these conclusions is that they are limited to the acute toxicity. Comparisons of the levels of acute toxicity and chronic toxicity have been made for *Daphnia magna* and some fish species. In a subsequent paper Slooff *et al.* [5] pointed out that the ratios between acute and chronic toxicity vary less than the ratios between the susceptibilities in acute tests. A factor of 100 usually covered this range, with the median lying in the order of 10, whereas ratios of more than 1000 between susceptibilities of different species in acute tests are not uncommon. Therefore adding 1 or 2 chronic tests to the acute tests mentioned above will probably result in an effective set of laboratory tests for the prediction of a NEL.

The *Daphnia magna* reproduction test is already available for this purpose and the same is the case for the early life stage test with fish which will soon be added to the OECD Test Guidelines.

I would like to use these results for a short analysis of the stepwise test system of the '6th amendment', the EC directive 79/831/EEC, concerning the notification of chemicals to be newly marketed. This directive requires at a low marketing level (1 ton/year) a base set of ecotoxicity data consisting of the results of acute tests with *Daphnia magna* and fish species. For this set of tests a first impression of the ecotoxicity can be expected, but not much more than that. With increasing marketing volume (10 tons/year) additional information can be asked for, starting with a *Daphnia magna* reproduction test and inhibition of algal growth. This information is compulsory (with escape clauses) at 100 tons/year. If the marketing volume increases further tests with additional species and longer-term fish tests can be required, thus revealing a rather reliable estimate of the toxicity of a chemical to most aquatic species. The concentration of a newly marketed chemical in the environment can, in a first approximation, be assumed to increase with increasing marketing volume (although probably not linearly).

In Fig. 6.2 the result of this system is shown in a somewhat qualitative picture. The upper drawn line represents the experimentally determined NEL. The lower drawn line indicates the NEL one could expect if more information would have been available. The broken line gives the environmental concentration as a function of the marketing volume.

When marketing starts the NEL for the environment is only known with great uncertainty, but the environmental concentration will be very low, resulting in a sufficient safety margin.

With increasing market volume this safety margin could decrease, but is known with more certainty. One might expect that by the time the environmental concentration approaches the NEL, the latter is known well enough to base a policy decision upon. In this way an acceptable balance between cost of testing and certainty about protection of the environmental quality is obtained.

So far I have tried to illustrate that the current practice of relatively simple aquatic toxicity tests — acute tests with 3 to 6 species and a few chronic tests — results in an acceptable insight in the environmental hazard of chemicals. Of course this preliminary conclusion is only possible thanks to several simplifications I have made. I would like to discuss three of them shortly.

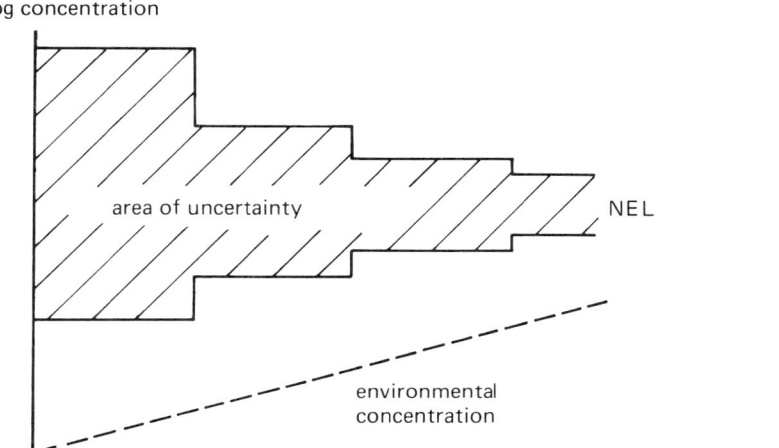

Fig. 6.2 — A justification of the step-wise testing system of the Directive 79/831/EEC (the '6th Amendment').

6.3 SPECIATION

The first one has a chemical background. The uptake of some chemicals, and therefore their toxicity, depends greatly on their actual form in the environment. This has been clearly shown for metals, where a different speciation, may result in an increase in their toxicity with 2 or even 3 orders of magnitude. The same holds for weak acids and bases, for which the toxicity depends on the environmental pH. Fortunately these influences are well known and their trend is predictable. If in the ecotoxicity tests the environmental conditions are chosen in such a way that they reflect the worst case, this is not a point of concern. It is, however, a point of particular interest when extrapolating from fresh water to salt water.

6.4 MULTISPECIES TESTS

The second one is that all the ecotoxicity tests discussed so far, are single species tests performed under optimal laboratory conditions, whereas the ecosystem consists of many species with complex interactions under environmental conditions which are more extreme than those in the laboratory. To narrow the gap between single species laboratory tests and real environment, multispecies tests, ranging from small-scale laboratory systems to semi-field trials are proposed.

Experimental results of multispecies tests, however, do not show no-effect-levels which are clearly lower than those in single species test [6]. Kooijman [7] has given several possible reasons for this phenomenon, which can be summarised as follows:
- Multispecies systems are not more susceptible to toxicants than each of the individual species. If this should be the case the current practice of single species testing is satisfactory.
- In multispecies systems mechanisms exist to compensate the stress of the toxicant. Some of these mechanisms are well known. Chemical fate plays a more important role in these tests than in single species tests. If, however, NELs are based on actual instead of on nominal concentrations this explanation is no longer applicable. Further, populations may be less sensitive, for example, because of selection of resistant individuals. The relatively long test durations gives these compensation mechanisms more chance. The quantitative importance of these mechanisms is, however, hard to assess.
- Effects do occur at lower levels, but because of the variation in the blank values of the test parameters, it is difficult to detect them. This is indeed one of the major troubles for the experimenter in this field. Many replicates would be necessary to reduce this noise in test parameters, but tests are too laborious and expensive to set up on a large scale.

Probably not one of these reasons alone will explain the frequently observed failure of multi-species tests to be more sensitive than single-species test. I consider, however, the last reason mentioned to be very important. Our present knowledge of ecological processes is too limited to decide if a change in one of the variables is caused by the toxicant or not. What is true for laboratory multi-species tests, certainly holds for field studies. It is certainly not my intention to say that multi-species tests including field-studies are of no value. More complex test systems, prefereably at field-level are indispensable when predictions about the degree of damage or recovery rates have to be made — a subject I excluded from this chapter. For predicting no-effect-levels too many difficulties still arise.

As a compromise between single and multi-species tests, Kooijman has performed experiments with populations of daphnids under chemical and food stress simultaneously. This type of test is much more complex than the usual toxicity test, but helps to analyse the ecologically important predator — prey relationship. The present stage of development of this type of testing, does not allow its routine application. The first experimental results do show, however, the importance of very low effect levels, which are sometimes considered to be irrelevant because of the severe effects of natural changes in field conditions. It is obvious from Kooijman's work that under conditions where a food stress or a chemicals stress alone have only a minor impact on the population growth rate of the wheel animal *Brachionus rubens* the combination of these stresses can be disastrous (Fig. 6.3).

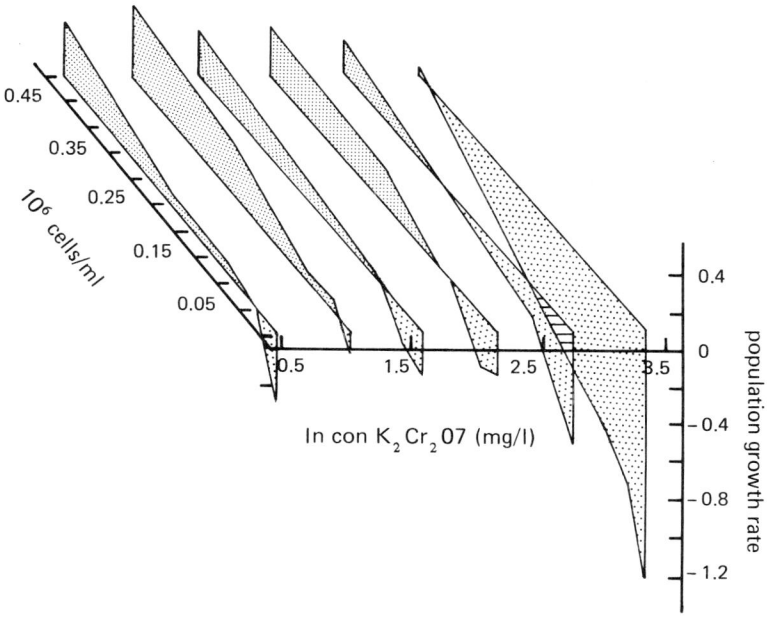

Fig. 6.3 — The simultaneous influence of a chemical and a food stress on the population growth rate of *Brachionus rubens*.

My conclusion on the relevance of multi-species testing in that at the present stage of its development a routine application can be excluded, yet the use of a safety factor for the extrapolation from single species tests to a multi-species system seems to be necessary.

6.5 TOXICITY OF MIXTURES

The third complication in the prediction of NELs for the environment is the simultaneous presence of many toxicants in the aquatic environment, a subject which I would like to discuss also because of my own scientific affinity to it. At first sight this complication seems very difficult to overcome. It is generally known that the toxicity of chemicals can show all kinds of interaction, which makes their combined toxicity difficult to predict. Both antagonism and potentiation are observed for the joint action of two toxicants. The environmentally relevant situation, however, is not one of mixtures of 2 chemicals but of tens, hundreds or thousands of chemicals.

The WRC is one of the few laboratories in which toxicity experiments with more complex mixtures have been performed, in which the environmental situation was mimicked (with up to 5 chemicals) [8,9].

In my own experiments [10] and in those of Hermens et al. [11] the attempt is made to go somewhat further, particularly by extending the number of chemicals and by choosing the concentrations on statistical grounds. There are two types of mixtures which have been investigated in these experiments: I will not burden you with the many obscure definitions in this field but give you the results of these experiments in a rather simple form.

1. *Mixtures of chemicals with a similar mode of action.*

    For these chemicals in all examples in LC50 tests with fish their toxicity was completely additive:

    $$\frac{C_1}{LC50_1} + \frac{C_2}{LC50_2} + \ldots + \frac{C_n}{LC50_n} = 1$$

    Table 6.2

    Examples of joint toxicity (LC50 guppy) of chemicals with a similar mode of action

Chemicals		
Type	Number	$\sum_{i=1}^{n} C_i/LC50_i$
Chlorobenzenes and toluenes, chloroalkanes and alkenes glycol derivatives a.o.	50	0.9
Chlorophenols	11	1.0
Anilines	17	1.1

    These chemicals were selected for similarity of action in the basis of quantitative structure — activity analyses (Hansch approach).

2. *Mixtures of chemicals with different modes of action.*

    Mixtures of 8 to 24 chemicals have been tested. In none to these cases antagonism or potentiation has been observed. In all cases the combined toxicity was slightly less than additive.

    Table 6.3

    Examples of joint toxicity (LC50 guppy) of chemicals with dissimilar modes of action

Number of chemicals in the mixture	$\sum_{i=1}^{n} C_i/LC50_i$
8	1.3
9	2.5
24	2.3

    Since these chemicals have been selected because of their dissimilarity in mode of action, also their chemical structures are very divergent and cannot be characterised in a few words.

It is likely that for more complex mixtures the situation will be the same. Potentiation or antagonism cannot be expected.

These results do not help us in establishing NELs for individual chemicals. Especially the results of experiments of chemicals with a similar mode of action indicate that even levels below the NELs may together result in significant toxicity. The only remedy I see for this problem is again taking it into account when choosing a safety factor and to consider it as a stimulus for ongoing efforts to reduce levels of toxicants in the environment, also when these are below their NELs.

6.6 CONCLUSIONS

The present state of the art of ecotoxicity testing allows a reasonable estimate of the level of a chemical which can be expected to have no negative impact on the environment. A set of about 5 acute tests and 2 chronic tests seems to be sufficient for this purpose. Safety factors are necessary to accommodate differences between the real environment and the laboratory, in particular to extrapolate to the multi-species situation and the simultaneous presence of toxicants. Further scientific research has to be done to support a choice for these safety factors.

6.7 REFERENCES

[1] OECD (1981) *Guidelines for the testing of chemicals.* OECD, Paris.
[2] Alexander, H. C., Bodner, K. M., and Mayes, M. E. (1983). *Chemosphere*, **12**, 415.
[3] McKim, J. M. (1977). *J. Fish. Res. Board Can.*, **34**, 1148.
[4] Slooff, W., Canton, J. H., and Hermens, J. L. M. (1983). *Aquatic Toxicology*, **4**, 113.
[5] Slooff, W., and Canton, J. H. (1983). *Aquatic Toxicology* (in press).
[6] Slooff, W. (1983) *Proceedings of ESA/SETAC Symposium on Multispecies Toxicity testing* (in press).
[7] Kooijman, S. A. L. M. (1983) *Proceedings of ESA/SETAC Symposium on Multispecies Toxicity Testing* (in press).
[8] Brown, V. M. (1968), *Water Res.*, **2**, 415.
[9] Brown, V. M., Shurben, D. G., and Shaw, D. (1970) *Water Res.*, **4**, 363.
[10] Könemann, H. (1981). *Toxicology*, **19**, 229.
[11] Hermens, J., and Leeuwangh, P. (1982). *Ecotoxicol. Environm. Safety*, **6**, 302.
[12] E. C., (1979) *Off. J. Eur. Comm.* L. 259/24.

Discussion

Chairman: **PROFESSOR T. R. E. SOUTHWOOD, FRS**
Chairman, Royal Commission on Environmental Pollution

Discussion began with a comment on Dr Könemann's presentation that one does find more than additive effects more commonly with pesticide mixtures than with the other commonly occurring substances in polluted rivers. Referring to the EIFAC report on mixtures it was noticed that for commonly occurring substances such as ammonia, phenol, cyanide and the heavy metals (especially Cu and Zn) the effects of combinations of these is approximately additive and that one can ignore fractions of the lethal values which are below the 'no-effect' level when considering the effect of mixtures.

Dr W. H. Könemann replied that there are certain combinations of chemicals which can show additive or antagonistic properties such as those used in pesticide formulations. However, the more complex the situation the less likely it is that these kinds of effects will be observed.

His feelings about the EIFAC paper were that the authors had done a good job considering the available information but there were very few experiments performed and in some instances the effects could be stronger than indicated in the report.

The point was made that even though the toxic effects may be additive, where the concentrations were so low as to be below the minimum effect for that individual compound then it could be discounted from the sum total. It was also important when setting environmental criteria to look at the two ends of the scale: the highest concentrations of toxic materials at which fish can survive and the lowest concentration at which an effect is observed.

The discussion continued with Dr P. Fraser's comparison of the falling mortality from stomach cancer over time with the increasing nitrate levels in water. It was pointed out that this was only evidence against nitrate being responsible for a large proportion of stomach cancer deaths. It would still be

Discussion

possible for nitrate to be responsible for a small number of the total stomach cancer deaths and for the small proportion to be increasing, while the majority of stomach cancers were decreasing with time.

Dr Fraser replied that this hypothesis assumed that nitrates were being reduced to nitrites, and then converted to nitrosamines which are carcinogenic in humans. However, this pathway has yet to be proved. An earlier speaker had highlighted the difficulty in extrapolating from species to species and from high dose to low dose and this was the situation for the N-nitrosamines. Although they appear to be potent carcinogens when given to animals, the relevance of this to humans is still debatable.

There was some concern over some of the epidemiological studies spoken about by Dr Fraser in her presentation. The concern was that when no effect had been found regarding epidemiological surveys, perhaps the study had insufficient power to show up a positive correlation. The example of Shirley Beresford's work in London where lowland surface water was not shown to be significantly more hazardous than groundwater in the London Area was quoted.

Dr Fraser was satisfied that Shirley Beresford's work did have sufficient power to provide meaningful results owing to the large population being studied.

A second concern was associated with the Worksop nitrate scare in the early 1970s. The water authority took action to minimise the use of the offending source, sank a new borehole, built a new pumping station and pipeline to replace it, and terminated the use of the Worksop Sewage Farm which was the main source of nitrate into the aquifer.

Although the decision was right based on the information available at the time would the same decision have been taken today. Dr Fraser thought that now, with the information we have the authority would probably not have sunk a new borehole. Termination of the use of the sewage farm was clearly sensible to avoid further pollution.

Referring to the problem of nitrates and stomach cancer, she said that, ideally, studies should be carried out on individuals where total nitrate intake had been measured, that is in populations at differing risk or in individuals with or without stomach cancer. The problem here is in knowing what their total nitrate intake was ten years or so ago, before they developed stomach cancer. Unfortunately, this information is not available.

Dr Fraser was asked whether the mobility of people over a large timescale reduced the power of an experiment. She replied that this was something that must be considered during an analysis and that it may be a source of bias.

There was discussion on the proportion of nitrate from drinking water in the total dietary intake, which was available from water. In the case of less than 50 mg/l it was 30%. However, at concentrations of greater than 100 mg/l the level rose to 83%. In the case of benzene (see Chapter 4) only 0.5% is attributed to the water supply. Since waterborne nitrate is such a large contributor to total intake, should we not still show some concern?

Dr Fraser replied that we are showing concern since we adhere to WHO guidelines concerning nitrate levels in the water supply. Because of this a large percentage of the population are not exposed to waterborne nitrate at levels greater than 50 mg/l. The small section of the public who are using well-water have their water supply screened during times of high risk, for example bottle-fed infants who are susceptible to methaemoglobinaemia.

Dr Fraser was asked whether nitrate intake included beer and whether the nitrate level correlation between beer and water had any influence on stomach cancers when comparing the incidence in males and females.

Dr Fraser replied that the well-users had been asked to keep a diet-diary over two consecutive weekends recording everything they ate or drank and this included beer consumed. However, the beer was not analysed for nitrate and so the nitrate contribution from beer was not known.

The point was made that at current waterborne nitrate levels methaemoglobinaemia is the main cause of concern regarding nitrate and not stomach cancer. Dr Fraser repeated her earlier comment on the uncertain link between nitrate and stomach cancer.

Delegates were reminded of a government-backed epidemiological survey looking into the effects of the nuclear tests in the early sixties on servicemen. A minister commented that the work had only started now since it takes approximately twenty years for the effects of the nuclear tests to become apparent. There was some suspicion that the same time lag may occur for nitrate-induced stomach carcinomas and since it is approximately twenty years since East Anglian nitrate levels rose sharply, perhaps Dr Fraser's work was somewhat premature.

Dr Fraser replied that in the East Anglian survey they had been looking at the mortality rate around the 1971 census in relation to nitrate levels going back to 1955. Data previous to 1955 were not available. Dr Fraser thought that a ten- to fifteen-year lag was reasonable in this field. The work would also provide an excellent background baseline for comparison with future surveys.

Part 3

PROTECTION AND CONTROL

CHAPTER 7

Standards and limits for water quality control

R. F. PACKHAM
Water Research Centre

7.1 INTRODUCTION

Water, whether processed or in its natural state, has more numerous diverse and important uses than any other material. Most of these uses are affected to some degree by its composition and in many cases 'water quality' is a vital consideration. Research into the effects of water quality often culminates at the practical level in the development of a concentration limit, above or below which some undesirable effect is manifested.

Such limits, although frequently criticised, are in fact in considerable demand by organisations and individuals needing some simple guidance in order to manage a particular part of the water cycle. While they are highly desirable for this purpose, the existence of a limit usually over-simplifies a highly complex situation. The need to understand fully the nature of the number representing the limit is obvious.

When, as frequently happens, limits become enshrined in national or international law they take on a new character. The importance of the number itself is sharpened; it becomes either a refuge or a threat depending on whether the limit can be complied with or not. The threatening aspects of a limit are real. Inability to comply can mean heavy expenditure on remedial measures or the closing down of an otherwise satisfactory resource. The validity of limits and the way in which they are to be applied becomes a crucial matter in this situation.

Drinking water standards have a long history and a discussion of these will illustrate many of the characteristics of limits for water constituents that are equally relevant to other types of water. The situation regarding limits for the protection of aquatic life will also be reviewed.

In both of these examples the limits are set to define the quality of water required for a particular purpose. There are however a number of EC Directives

7.2 DRINKING WATER STANDARDS

7.2.1 Characteristics

Drinking water standards have been developed to define a quality of water that is safe and acceptable to the consumer. It would be inappropriate to standardise the composition of drinking water in a comprehensive way; the primary purpose of standards is to provide guidance on health related aspects of water quality that are outside the normal competence of those responsible for providing public water supplies. Most drinking water standards therefore set limits for organisms or chemical substances that are dangerous, potenially hazardous or obnoxious to consumers. Obnoxious materials include those that though harmless to health, may have an adverse effect on the use of the water due to taste, odour, colour, turbidity, corrosivity or other undesirable properties.

Drinking water standards of the United States Public Health Service appeared as early as 1914 [1], while those of the World Health Organization (WHO) were first published in 1958, with later editions in 1961, 1962, 1970 and 1971 [2–6]. Many countries have national standards and a substantial number of these are based on the WHO International Standards for Drinking Water. The WHO International and European Standards have now been combined and completely revised as WHO Guidelines for Drinking Water Quality [7] to be published in 1983. The European Community has published a Directive on the Quality of Water for Human Consumption [8] which will be fully implemented in the United Kingdom in 1985.

In all published drinking water standards the contaminants or constituents of interest are subdivided in various ways to cover the following general categories:

(i) Microbiological and biological quality.
(ii) Chemical constituents of possible health significance.
(iii) Chemical constituents related to aesthetic and other considerations.
(iv) Radiological aspects of water quality.

Discussion will be limited here to the first three of these categories.

7.2.2 Microbiological and biological quality

The greatest potential health hazard associated with the public water supply is the possibility that it could become a vehicle for the spread of a waterborne infectious disease. Although the widespread disinfection of water supplies has virtually eliminated this problem in developed countries, waterborne bacterial diseases such as cholera, typhoid and paratyphoid account for the death of millions annually in the Third World. The basic problem is a lack of, or inadequate, disinfection of a water supply which is contaminated with faecal material of either human or animal origin and which may therefore contain viable causal agents of enteric disease.

Although methods are available for the detection in water of the pathogenic organisms of concern, it is far more practicable to evaluate microbiological quality in terms of the more easily detectable normal intestinal organisms which vastly outnumber any pathogens likely to be present in faeces. The organisms most commonly used as indicators of faecal pollution are *Escherichia coli* and the coliform group as a whole.

Standards differ in relation to the precise definition of the primary bacterial indicator of faecal pollution (*E. Coli* or 'faecal coliforms') but all require these organisms to be absent from water supplies. The validity of this approach which has been highly successful in controlling waterborne disease in most developed countries, has been amply justified. There are differences between standards in relation to the acceptable level of contamination by organisms that are not specifically of faecal origin and these can be of operational importance.

There has been considerable discussion as to whether water supplies treated to normal bacteriological standards are adequately protected from waterborne viral pathogens. The most important of these is infectious hepatitis for which there are a number of undoubtable waterborne epidemics on record. A careful examination of these epidemics has shown, however, that in all cases massive contamination of the water supply or disinfection failure either occurred or could not be ruled out. There is therefore no basis for setting a special limit for viruses in drinking water and this is indeed fortunate in view of the inadequacy of the techniques currently available for virological examination.

7.2.3 Chemical constituents

In early drinking water standards microbiological aspects of water quality were clearly the main consideration. Although this should still be true, some recent standards fail to give this impression. Table 7.1 shows the development of limits for chemical constituents in the drinking water standards of the US Public Health Service [1, 9–12]. These standards clearly provided the foundation for the drinking water standards of the World Health Organization from which in turn many national standards have been derived. Table 7.2 shows the growth in the number of limits for chemical constituents put forward by various standard setting organisations. Initially, limits were set only for inorganic constituents but since 1970 there is evidence of growing interest in the significance to health of trace organic constituents.

The precise reason for the inclusion of some substances that always appear in drinking water standards is uncertain, but it may stem from anxiety rather than from firm evidence of harmful effects at levels likely to occur in water. The setting of limits has undoubtedly been stimulated by the development of analytical

Table 7.1

US Public Health Service Drinking Water Standards (1914–1962).
Concentrations in ppm (1925–1956) and mg/l (1962).

Notes
1. The 1914 edition set bacteriological limits only.
2. In the standards for 1942, 1946 and 1962 it is stated that, for those constituents marked with an asterisk, levels in excess of the limit constitute grounds for rejection of the supply. Other limits should not be exceeded where other more suitable supplies are available.

Year	1925	1942	1946 1956	1962
Alkylbenzene sulphonates (ABS)		0.5		
Arsenic		0.05*	0.05*	0.05* (0.01)
Barium				1*
Cadmium				0.01*
Carbon chloroform extract (CCE)				0.2
Chloride	250	250	250	250
Chromium VI			0.05*	0.05*
Colour	20			15
Copper	0.2	3.0	3.0	1
Cyanide				0.2* (0.01)
Fluoride		1.0	1.5	0.6–1.7†
Iron	0.3			0.3
Iron + Manganese		0.3	0.3	
Lead	0.1	0.1*	0.1*	0.05*
Magnesium	100	125	125	
Manganese				0.05
Nitrate				45
Phenolic compounds		0.001	0.001	0.001
Selenium		0.05*	0.05*	0.01*
Silver				0.05*
Sulphate	250	250	250	250
Total solids	1000	1000	1000	500
Turbidity	10			5
Zinc	5.0	15	15	5

† Depending on temperature.

Table 7.2

Limits for drinking water constituents of concern in relation to health

Organisation	Ref.	Year	Number of limits set		
			Inorganic	Organic	Total
US Public Health Service	(10)	1942	3	0	3
	(11)	1946	4	0	4
	(12)	1962	8	0	8
World Health Organization	(2)	1958	5	0	5
	(3)	1961	6	0	6
	(4)	1963	9	0	9
	(5)	1970	8	1	9
	(6)	1971	8	1	9
	(7)	1983	9	18	27
US Environmental Protection Agency	(13)	1975	9	6	15
Health and Welfare Canada	(14)	1978	15	17	32
European Community	(8)	1980	9	2	11

techniques which allow the detection of substances of possible concern to be revealed at very low concentrations. The list of limits is likely to continue to grow for several reasons:

(i) Industry continues to develop new chemicals or new uses of existing chemicals at a rapid rate. The use of industrial chemicals may result in the contamination of a water source.

(ii) There are important gaps in the capability of existing techniques for the analysis of water. These gaps will eventually be filled and the presence of new substances of possible concern will be detected.

(iii) Health effects research may indicate a need to control the concentration of certain water constituents.

The large number of limits for chemical substances does present problems in relation to the feasibility and cost of monitoring particularly where there is a legal requirement to monitor for compliance. This aspect is dealt with more fully in Chapter 8. Although there is no doubt that all water supplies should be regularly monitored for bacteriological contamination the same is not true for all chemical contaminants many of which can be identified clearly with specific circumstances which do not arise in all supplies, that is lead from lead pipes and tanks, cyanide and chromium from industrial effluents, high pesticide

Ch. 7]   **Standards and Limits for Water Quality Control**   113

levels with their use in vector control, arsenic with certain groundwaters. It is doubtful whether routine monitoring of a supply for such substances can be justified unless there is evidence that the supply is at risk.

7.2.4 Setting the limits

Judged relative to any other kind of food the chemical content of drinking water is minute and there are very few substances for which the intake from water represents a significant proportion of the total dietary intake. Because of their low concentration the chemical components of drinking water are unlikely to cause effects on health in the short term. Any such effects that exist are more likely to result from intake over a long period of time. In setting limits for such substances acute toxicity data is usually almost irrelevant but comprehensive information on chronic toxicity and carcinogenicity is highly desirable. In considering such information is is necessary to identify any population groups particularly at risk such as infants, pregnant women, or other consumers with specific physiological defects.

Information is also required on the normal intake of the substance from different routes of exposure, that is, air, food and water. On the basis of this it is necessary to make a judgement as to the acceptable daily intake of the substance and its apportionment between the various routes of exposure and in particular to water. It is necessary also to apply a safety factor, the size of which will be determined by the strength of the toxicological evidence and the nature and magnitude of any health risk. The final judgement to be made is of course the magnitude of the limit itself.

As discussed in Chapter 4, the information available on the health effects and the routes of human exposure of water constituents is invariably inadequate. There is an almost total lack of hard evidence for health effects of chemicals at the concentration levels found in water and for several reasons this is not surprising.

Health effects associated with submicrogram per litre concentrations of organic compounds in drinking water, if they exist at all, are likely to be very small and to be manifested on the incidence of diseases such as cancer that are already widespread in the population. In most epidemiological studies such small increases in disease incidence due to drinking water are very difficult to perceive against a normal background incidence due to other more important factors. The considerable induction period (20 to 30 years) for cancer also creates enormous problems in trying to link cause and effect particularly as exposure information relating to 30 years ago for water constituents of interest today is almost non-existent.

The toxicological information available for water constituents is also defective in many respects and normally relates to levels of exposure that are orders of magnitude greater than those experienced from drinking water. Such data also invariably relates to tests with laboratory animals rather than man himself. The necessity to extrapolate from data resulting from high exposures with laboratory animals to very low exposures with man leads to considerable difficulty in assigning a limit to some water constituents.

The precautionary nature of some limits is emphasised where these are set, not on the basis of direct toxicological evidence at all, but on the minimum concentration detectable by analysis (for example chromium [11], cadmium [6]). Although such limits may seem strange they are preferable to those set *below* the limit of detection. There were several examples of this, for example, phenols [10] and pesticides [8].

7.2.5 Types of limit

The difficulties and uncertainties involved in setting limits for drinking water constituents are especially great with carcinogenic substances and the best available extrapolation techniques for these substances can give a limit which could be out by an order of magnitude either way [7]. The specification of limits in such widely used absolute terms as 'Maximum Permissible Concentration' and 'Maximum Acceptable Level' is inappropriate, and this has been brought out clearly in WHO Guidelines for Drinking Water Quality in which the limits have been set in terms of Guideline Values. The nature of these is defined as follows:

(a) A Guideline Value represents a concentration or number which ensures an aesthetically pleasing water and does not result in any significant risk to health of the consumer.

(b) The quality of water defined by the Guidelines for Drinking Water Quality is such that it is suitable for human consumption and for all usual domestic purposes including personal hygiene. However water of a higher quality may be required for some special purposes such as renal dialysis.

(c) A Guideline Value is to be used as a signal:
   (i) to investigate the cause when values are exceeded with a view to taking remedial action,
   (ii) to consult with authorities responsible for public health advice.

(d) Although the Guideline Values describe a quality of water acceptable for lifelong consumption, the establishment of these guidelines should not be regarded as implying that the quality of drinking water may be degraded to the recommended level. Indeed, a continuous effort should be made to maintain drinking water quality at the highest level of purity.

(e) The Guideline Values specified have been derived to safeguard health on the basis of lifelong consumption. Short-term exposures to higher levels of chemical constituents such as might occur following an accidental spill, may be tolerated but need to be assessed on a case-by-case basis, taking into account, for example, the acute toxicity of the substance involved.

(f) Short-term excursions above the Guideline Values do not, necessarily, imply that the water is unsuitable for consumption. The amount by which, and the duration for which, any Guideline Value can be exceeded without affecting public health depends on the specific substance involved.
    It is recommended that when a Guidelines Value is exceeded, the surveillance agency (usually the authority responsible for public health) be consulted for advice on suitable action based upon considerations such as the intake of the substance from sources other than drinking-water (for chemical constituents), the likelihood of adverse effects, the practicality of remedial measures and other similar factors.
(g) In developing national drinking water standards based on the WHO Guidelines it will be necessary to take account of a variety of local, geographical, socio-economic, dietary and industrial conditions. This may lead to national standards that differ appreciably from the Guideline Values.
(h) In the case of radioactive substances, the term Guideline Value is used in the sense of 'Reference Level' as defined by the International Commission on Radiological Protection (ICRP).

7.2.6 Present status

Everyone wants drinking water having all the desirable characteristics embodied in the word 'wholesome' that for so long provided the only legal quality requirement of water supplies in the United Kingdom. It is clear that many organisations believe that wholesomeness can be achieved only through standards which set out quality requirements in terms of rather stark limits, a growing number of which are for chemical constituents. Table 7.3 gives a complete list of the chemical substances for which limits have been set since 1970.

Table 7.3
List of drinking water constituents for which limits have been set

Parameter	Reference
Aesthetic organoleptic and physicochemical	
Colour	C D E F
Odour	C E F
pH	C D F
Taste	D E
Temperature (C)	D E
Turbidity	B C D E F

Table 7.3 – *continued*

Parameter						
			Reference			
Inorganic						
Aluminium					E	F
Ammonia	A				E	
Antimony					E	
Arsenic	A	B		D	E	F
Barium		B		D		
Boron				D		
Cadmium	A	B		D	E	F
Carbon Dioxide, Free	A					
Chloride	A		C	D		F
Chromium	A	B		D	E	F
Copper	A		C	D		F
Cyanide	A			D	E	F
Dissolved Solids, Total				D		F
Dry Residue					E	
Fluoride	A				E	F
Hardness, Total	A					F
Hydrogen Sulphide	A					
Iron	A		C	D	E	F
Lead	A	B		D	E	F
Magnesium	A				E	
Manganese	A		C	D	E	F
Mercury		B		D	E	F
Nickel					E	
Nitrate	A	B		D	E	F
Nitrite				D	E	
Nitrogen, Kjeldahl					E	
Oxygen, Dissolved	A					
Potassium					E	
Selenium	A	B		D	E	F
Silver		B		D	E	
Sodium					E	F
Sulphate	A		C	D	E	F
Sulphide				D		
Uranium				D		
Zinc	A		C	D		F

Standards and Limits for Water Quality Control

Table 7.3 – *continued*

Parameter	A	B	C	D	E	F
Organic						
2,4-D		B		D		F
1,1-Dichloroethane						F
1,2-Dichloroethane						F
2,4,5-TP		B		D		
2,4,6-Trichlorophenol						F
Aldrin + Dieldrin				D		F
Benzene						F
Benzo (a) Pyrene						F
Carbaryl				D		
Carbon Tetrachloride						F
Chlordane				D		F
Chloroform						F
Chloroform Extractables	A					
DDT				D		F
Detergents, Anionic	A				E	
Diazinon				D		
Endrin		B		D		
Heptachlor + Heptochlor Epoxide				D		F
Hexachlorobenzene						F
Hydrocarbons, Dissolved or Emulsified					E	
Lindane		B		D		F
Methoxychlor		B		D		F
Methyl Parathion				D		
Nitrilotriacetic Acid				D		
Oxidizability (KMnO4)					E	
Parathion				D		
Pentachlorophenol						F
Pesticides, Total	A			D	E	
Phenols and Phenolic Compounds	A			D	E	
Polycyclic Aromatic Compounds	A				E	
Tetrachloroethylene						F
Trichloroethylene						F
Trihalomethanes				D		
Toxaphene		B		D		

A World Health Organization European Standards for Drinking Water 1970 (5).
B US Environmental Protection Agency Primary Regulations 1977 (13).
C US Environmental Protection Agency Secondary Regulations 1979 (15).
D Health and Welfare, Canada 1978 (14).
E Council of the European Communities 1980 (8).
F World Health Organization, Guidelines 1983 (7).

Decisions on the levels chosen for these limits, most often have to be taken in advance of the necessary science and involve considerable professional judgement. Uncertainties are considerable but it is likely in most, if not all, cases that the levels are on the safe side. Information on the basis of each limit is vital as this may need to be put in the balance when decisions are to be made regarding any measures necessary to achieve compliance which involve substantial expenditure.

There has been concern for example about levels of nitrate in excess of the EC Maximum Acceptable Concentration in some English water supplies. There have been no cases in the United Kingdom of infantile methaemaglobinaemia, the disease on the basis of which the limit was set, for several years and only ten cases altogether in about 25 years. There are undoubtedly other factors as well as water nitrate involved in this disease and there is evidence that malnutrition is important. It has been suggested that high nitrate levels may be associated with certain forms of cancer but as fully discussed in Chapter 5 this is not borne out by any tangible evidence in the UK. A very careful examination of the facts would clearly be necessary before embarking on the removal of nitrate from water which would involve substantial costs.

Following a review of the evidence on which an EPA limit for trihalomethanes (THM) in water was based the DOE/DHSS Committee on Medical Aspects of Water Quality advised that although the removal of THM from water was not a matter of high priority, it would be prudent to take steps to reduce the concentration of THM as far as practicable in the light of other calls on resources and while retaining basic water disinfection techniques.

By way of contrast the evidence concerning the toxicity of lead and the important contribution of water lead to the total intake in some parts of the United Kingdom has not been questioned and a major programme aimed at reducing water lead levels is indeed under way.

These examples illustrate the need to consider cases of non-compliance on their individual merits. This contrasts sharply with the approach of USPHS and some early WHO drinking water standards in which levels in excess of the concentration listed constituted grounds for the rejection of the supply (see Table 7.1).

The more pragmatic use of limits is clearly embodied in the WHO Guidelines for Drinking Water Quality, both in the concept of the Guideline Value and in the provision of health criteria including in some cases information relating to constituents for which limits were not set (for example, asbestos, barium, hardness, sodium). The standards set by the Environmental Protection Agency and Health Welfare Canada are also supported by excellent sections dealing with the basis of the limits.

The EC Directive on the Quality of Water of Human Consumption is by comparison a disappointing document which includes no justification for any of the Maximum Acceptable Concentrations that are set. Although some are clearly derived from the WHO European Standards of 1970 others including the limits set for sodium, potassium, magnesium, nickel, antimony and pesticides really require a full explanantion. The justification of limits is particularly important in a document that is destined to become a part of Community legislation.

7.3 WATER QUALITY FOR AQUATIC LIFE

It is highly desirable for many reasons that surface fresh water should be able to support healthy populations of fish and other forms of aquatic life. Limits set for chemical substances in such waters on the basis of their potential effect on aquatic life have, however, a much shorter history than is the case for drinking water constituents. A considerable amount of work by the US Environmental Protection Agency and its predecessors on the effect of water composition on biota has provided the foundation for national regulations in many countries. In Europe the European Inland Fisheries Advisory Commission (EIFAC) has been active in this field and has produced a series of reports on water quality criteria for European freshwater fish [16]. These criteria seem to have been drawn on in the drafting of an EC Directive in 1978 [17] on the quality of fresh waters needing protection or improvement in order to support fish life.

In developing standards for aquatic life there would appear to be the advantage that instead of having to extrapolate toxicity data from one species to another (for example, rat to man) it should be possible to obtain data for the species of interest itself. Unfortunately most information available on the effect of pollutants on aquatic life is limited to certain much studied species which have been used principally because of their ready availability. The rainbow trout, for example, is easily bred and is very sensitive to low concentration of certain pollutants of interest. Decisions based on trout data may perhaps err on the safe side but it is in no sense a 'typical' species of fish. In general cyprinid fish are less sensitive and more variable in their response to pollutants than salmonid species. Information available on the toxicity of pollutants to invertebrates is often almost entirely based on work with *Daphnia*. This easily bred creature is useful but hardly representative of the vast range of organisms at its trophic level.

Other problems with fish toxicity data are as follows:

(a) More than 95% of published data relates to acute toxicity (usually 96 hr), although for many substances chronic toxicity data would be more appropriate. The derivation of chronic toxicity from acute data although widely practised is highly controversial.
(b) Most toxicological work has been undertaken on adult fish although early life stages are more sensitive to many pollutants.
(c) Although it is known that some other aspects of water quality (for example, hardness, salinity) can have a dramatic effect on the toxicity of certain pollutants, these are often ill-defined or incompletely studied.

(d) There is a general lack of information for the marine environment on toxicity to fish although data on invertebrates is more extensive than in the freshwater environment.

(e) There is a serious lack of information relating the results of laboratory toxicity studies to field studies of fisheries.

Thus much of the available information on the toxicity of water constituents to aquatic life is subject to considerable uncertainties although many of these should succumb to further research. So far EIFAC has published tentative water quality criteria for finely divided solids, extreme pH values, temperature, ammonia, monohydric phenols, dissolved oxygen, chlorine, zinc, copper, cadmium, and chromium. Criteria for nickel, aluminium and nitrite are in preparation. The EIFAC reports are excellent, well documented and explanatory reports.

The EC Directive appears to have been partly based on the EIFAC reports but other considerations were introduced which have been commented on in WRC 'Notes on Water Research' [18]. In particular, limits are given in the Directive from some parameters which have little if any bearing on fish life (for example, BOD, total phosphorus), mandatory limits are set for some determinands subject to considerable natural variability (for example, temperature, dissolved oxygen) while the stringency of some other limits (for example, ammonia, nitrite) cannot be support by available data. On this Directive the EC Economic and Social Committee commented [19]

'... in several countries (of the EC) there are rivers that contain good fish populations in spite of the water quality being lower than specified in the Annex (of the Directive). The Committee, therefore, feels that the values in the Annex should be reconciled with reality'.

7.4 DISCUSSION

Despite all of the uncertainties involved in setting limits for water constituents, they are useful and even vital in many aspects of water quality management. The number of limits established for various purposes will continue to grow and some of the present uncertainties will diminish as the result of research.

Problems in urgent need of solution are the relative toxicity of different chemical species of the same element and the toxicity of mixtures of substances. EIFAC has proposed a method of dealing with the toxicity of mixtures of pollutants to aquatic organisms but at the present time there is no generally accepted approach that is considered appropriate for drinking water. More information is urgently required on the magnitude of the risk presented by trace quantities of potentially harmful substances in drinking water.

There can be no question as to the value of limits in providing guidance but their over-rigid application on a pass or fail basis can lead to problems. One wonders on how many occasions a concentration of a toxic substance in excess of the old USPHS or WHO drinking water standards has led to rejection of the supply (see Table 7.1) and what were the circumstances. There are certainly those who have said that in developing countries, WHO drinking water standards have done more harm than good by leading to the rejection of an otherwise satisfactory supply on the basis of, for example, a nitrite concentration slightly above the limit. The approach embodied in the new WHO Guidelines for Drinking Water Quality shortly to be published, offers a rational way of dealing with such difficulties. Values in excess of the Guideline Value trigger not the closing down of the supply, but an investigation into all aspects of the contamination including its significance to health in the local context (see section 7.2.5). There is ample evidence that the safety margins built into Guideline Values and most other commonly used limits for chemical constituents of drinking water are more than adequate for such an approach.

It is vital that these aspects of water quality standards, limits and guidelines should be clear in the minds of those who for any reason wish to embody these numbers in law. The need to safeguard public health and environmental quality at all times is unquestionable but excessive safety margins have important cost implications which cannot be ignored. Experience indicates that it takes a long time to get a bad law rewritten. Bad laws also fall into disrepute; loopholes are fully exploited and they are interpreted in a way that renders them largely ineffectual. Some at least of the EC water directives unfortunately seem destined for this treatment.

7.5 REFERENCES

[1] United States Treasury Department. (1914) Bacteriological Standards for Drinking Water. *Public Health Report*, **29**, 2959–2966.
[2] World Health Organization. (1958) *International Standards for Drinking Water.* WHO, Geneva.
[3] World Health Organization. (1961) *European Standards for Drinking Water.* WHO, Geneva.
[4] World Health Organization. (1963) *International Standards for Drinking Water.* WHO, Geneva.
[5] World Health Organization. (1970) *European Standards for Drinking Water.* WHO, Geneva.
[6] World Health Organization. (1971) *International Standards for Drinking Water.* WHO, Geneva;
[7] World Health Organization. *Guidelines for Drinking Water Quality.* WHO, Geneva, (in press).
[8] Council of the European Communities. (1980) Council Directive of 15 July 1980 relating to the quality of water intended for human consumption (80/778/EEC). *Official Journal of the European Communities*, No. L229/11, August 30.

[9] US Public Health Service. (1925) Report of Advisory Committee on Official Water Standards. *Public Health Report*, **40**, pp. 693–721.
[10] United States Public Health Service. (1943) *Public Health Service Drinking Water Standards 1942.* US Gov. Printing Office, Washington D.C.
[11] US Department of Health, Education and Welfare. (1946) *Public Health Service Drinking Water Standards 1946. Public Health Report*, **61**, pp. 371–384 (Reissued March 1956).
[12] US Department of Health Education and Welfare. (1962) *Public Health Service Drinking Water Standards 1962.* US Gov. Printing Office, Washington D.C.
[13] US Environmental Protection Agency. (1976) *National Interim Primary Drinking Water Regulations*, EDA-570/9-76-003. US Gov. Printing office, Washington D.C.
[14] Ministry of National Health & Welfare. (1978) *Guidelines for Canadian Drinking Water Quality*, 1978. Canadian Gov. Publishing Centre, Quebec.
[15] US Environmental Protection Agency. (1979) National Secondary Drinking Water Regulations, *Federal Register*, **44**, (140), 42195–42202, July 19.
[16] Alabaster, J. S. and Lloyd, R. (1982) *Water Quality Criteria for Freshwater Fish*. Second edn. Butterworths, London.
[17] Council of the European Communities. (1978) Council Directive of 18 July 1978 on the quality of fresh waters needing protection or improvement in order to support fish life. 78/659/EEC, Brussels.
[18] Water Research Centre. (1979) New European standards for freshwater fish. *Notes on Water Research* No. 20.
[19] European Economic Community. (1977) *Opinion of the Economic and Social Committee on the Proposal for a Council Directive on the quality requirements for waters capable of supporting freshwater fish.* CES 235/77 mb. Brussels, 2 February.

CHAPTER 8

Monitoring compliance with standards

J. C. ELLIS and D. G. MILLER
Water Research Centre

8.1 INTRODUCTION

With the move towards increasing legislation and the tendency to prefer a more rigorous numerical approach to standards, operating agencies are being obliged to examine in more detail the implications of framing and policing a variety of measures covering the whole field of environmental protection. This chapter reviews the statistical principles involved in compliance monitoring, and sets out plainly the compromises necessary in practice.

8.1.1 Definitions

The nature and derivation of *standards* have been explored in detail in the previous chapter. This ground will not be repeated here, but later in the chapter it will be necessary to consider some of the implications of different forms of standard. *Monitoring* can be defined as the measurement of quality, continuously or, more commonly, intermittently, for comparison with standards. This is examined in more detail later. *Compliance* could be defined as a successful outcome to the monitoring exercise, but the judgement of success or failure is easier said than done.

8.1.2 The ideal concept

Most people's instinctive concept of compliance is unrealistically clear-cut. In some cases this is due to a lack of awareness of the complexities involved. In others, perhaps where pressure groups and campaigners are concerned, each infringement is a cause, and a black and white approach helps the cause. This ideal situation, where quality is continuously measured and the standard is a straightforward limit never to be exceeded, is illustrated in Fig. 8.1. Rarely are such conditions met in environmental matters any more than they are in day-to-day affairs; even the classic case of speeding offences can sometimes be more complicated than it appears at first sight.

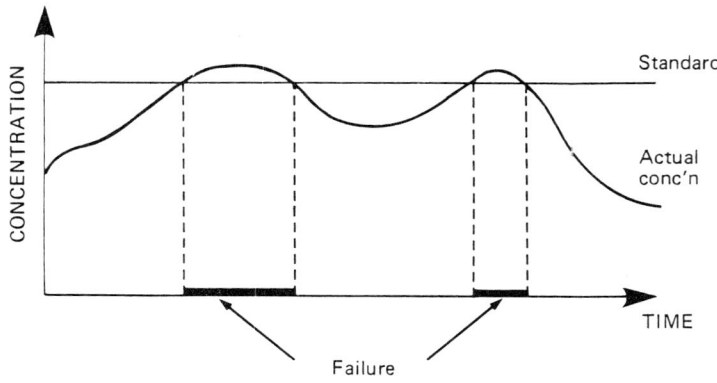

Fig. 8.1 — Compliance-testing in an ideal world.

8.2 THE REAL WORLD

Complications introduced in real-life situations include:

(a) quality variations which are far from smooth and often unpredictable;
(b) the more complex nature of standards;
(c) severe limitations on sampling effort, leading to time and space restrictions;
(d) the necessity of dealing with more than one determinand;
(e) errors in sampling and analysis.

8.2.1 Quality variations

Figure 8.2 shows examples of quality variation experienced in (i) river water; (ii) the final effluent from a sewage works; and (iii) potable water in a supply main. To devise a realistic policing policy in these different situations often requires the stark choice between delaying a massive sampling effort or tolerating a low probability of detecting serious deteriorations in quality. In some cases statistical analysis can assist in defining systematic patterns, thereby lessening the sampling effort required. Figure 8.3 shows the application of such an approach to river quality variations over time. In this case the existence of seasonal patterns enables sampling to be designed more cost-effectively. In Fig. 8.4, a similar approach is used to examine changes along a river, thus enabling sampling to be directed towards reaches where real changes occur (usually as the result of effluent discharges, tributaries or river structures). Statistical techniques of this kind will, of course, be successful only where substantial real patterns exist, and so there will be cases where the benefits are limited.

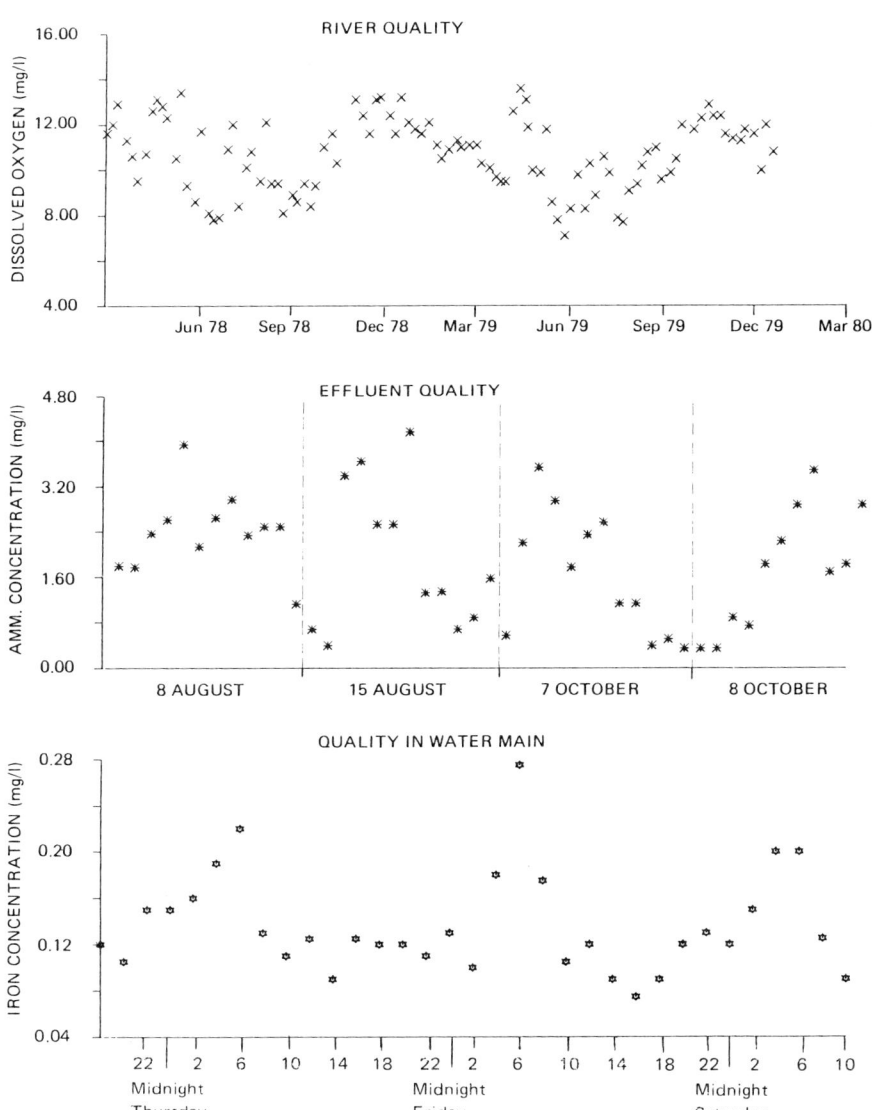

Fig. 8.2 — Types of quality variation.

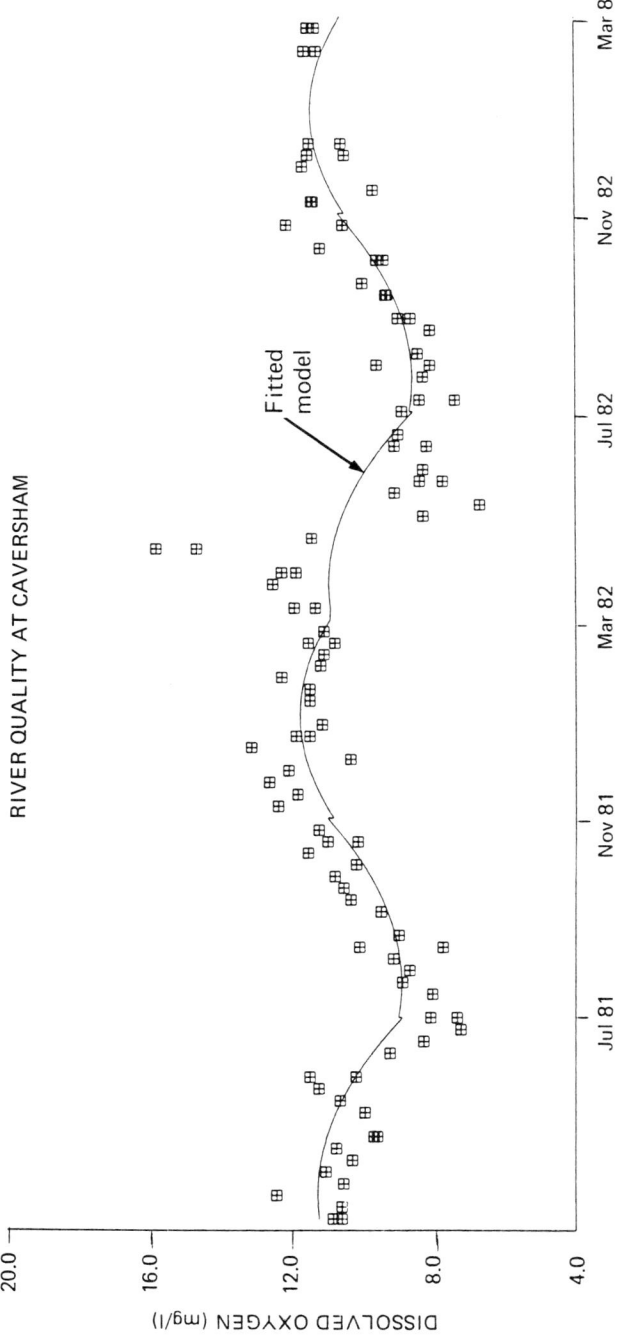

Fig. 8.3 – River quality variation through time.

Ch. 8]   Monitoring Compliance with Standards   127

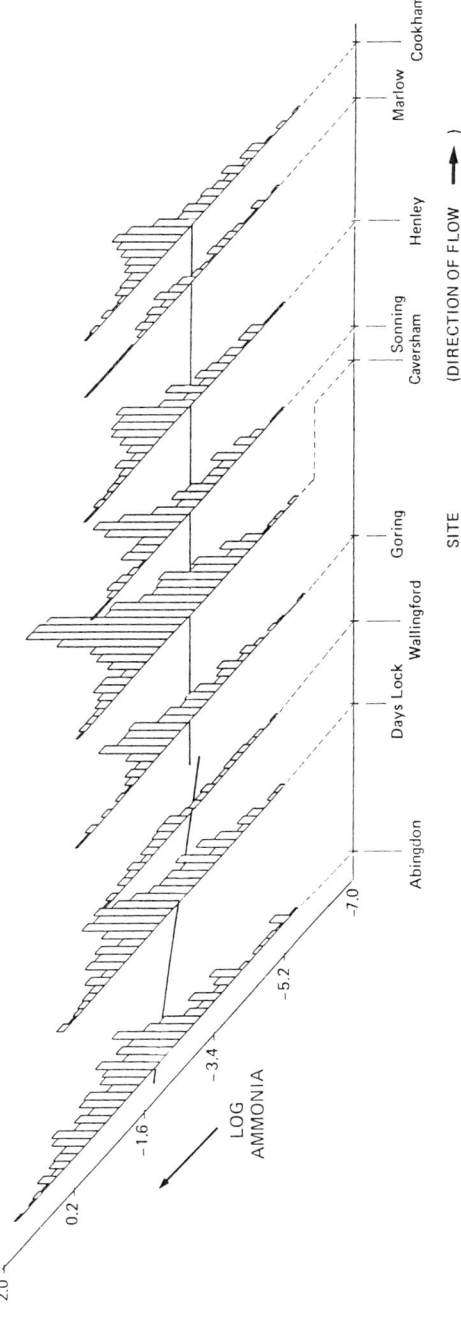

Fig. 8.4 — Spatial variation in river quality.

8.2.2 Standards

The previous chapter showed that rarely are standards derived from precise scientific data. Often values are based on inadequate data, and additionally may be strongly influenced by political compromise at the negotiating stage. Some standards are declared to be absolute concentrations beyond which sampled levels may not stray, but often standards are defined as mean or percentile concentrations, as illustrated in Fig. 8.5. A further factor is that standards are not always precisely defined and powers of interpretation are left with national or local agencies. It is a poor reflection, in fact, on the machinery of legislation where 'woolliness' is sometimes the saviour of the negotiator.

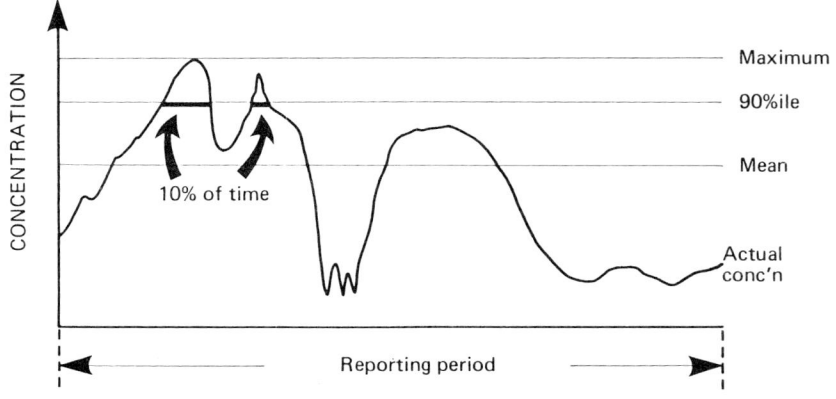

Fig. 8.5 — Types of standard.

Leaving aside the idiosyncracies of international negotiation, the type of standard may reflect the degree of seriousness of transgression. 'Maximum Admissible Concentration' implies that the value should not be breached at any cost. In practice, rarely is this the case since standards are seldom set at threshold levels beyond which instant effects set in. The percentile approach openly admits to some transgression on a percentage of occasions or in a percentage of cases, but must be handled with care. Figure 8.6(b) shows an instance where a 95 percentile compliance is achieved but the single excursion shown could be lethal to some environmental species. The mean seems more relaxed still, implying an interest solely in long-term accumulations. However, the time scale is important. For example, a mean to be complied with *daily* could actually be much more stringent than a 99 percentile standard to be complied with *annually*.

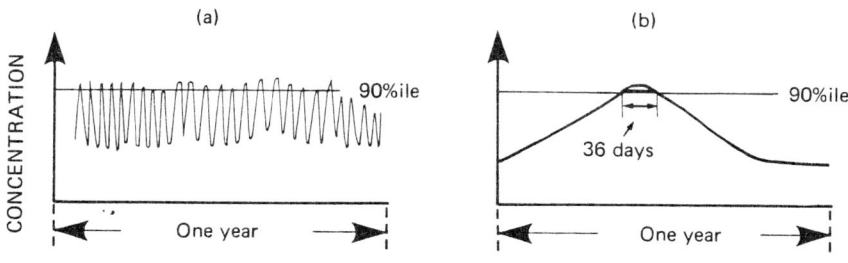

Fig. 8.6 — Two ways of achieving 90% compliance.

8.2.3 Sampling constraints

In the real world, only limited effort is available for policing operations, and operating agencies need to assess priorities against a background of constraints in manpower and other resources. The public at large, and pressure groups in particular, may not be aware of or ready to accept such limitations. These constraints heighten the need for a cost-effective approach, and this is dealt with more fully in section 8.3.

8.2.4 Multiple determinands

Where a number of determinands are involved in a regulation, the problem could become more complex and remedial measures much more expensive if a 'one-out-all-out' approach were taken towards compliance. If the standards are based on rigid MACs then there is little alternative to this approach and the solution is clear-cut. However, where a percentile standard is used and the failure of a *sample* is judged on the basis of any value of *any one determinand* in that sample exceeding the limit, then it can be shown statistically that there is a progressively increasing risk of failure as the number of determinands increases. Figure 8.7 illustrates this point. Multi-determinand compliance is the subject of a recent WRC report [1], which concludes that each determinand should be judged separately on a percentile basis, thus more truly reflecting the intention of the standard. That approach has been assumed in the remainder of this chapter.

8.2.5 Error of measurement

In considering compliance monitoring it is easy to forget that measurement of concentration is itself subject to error, as is the process of sampling and sample preservation. Errors may be random or systematic or both, and each will have its own impact on the reliability of surveillance. Adequate attention should be paid to such errors and their assessment, both to quantify the effects and to control them to an acceptable level in relation to the standard. With the tendency of

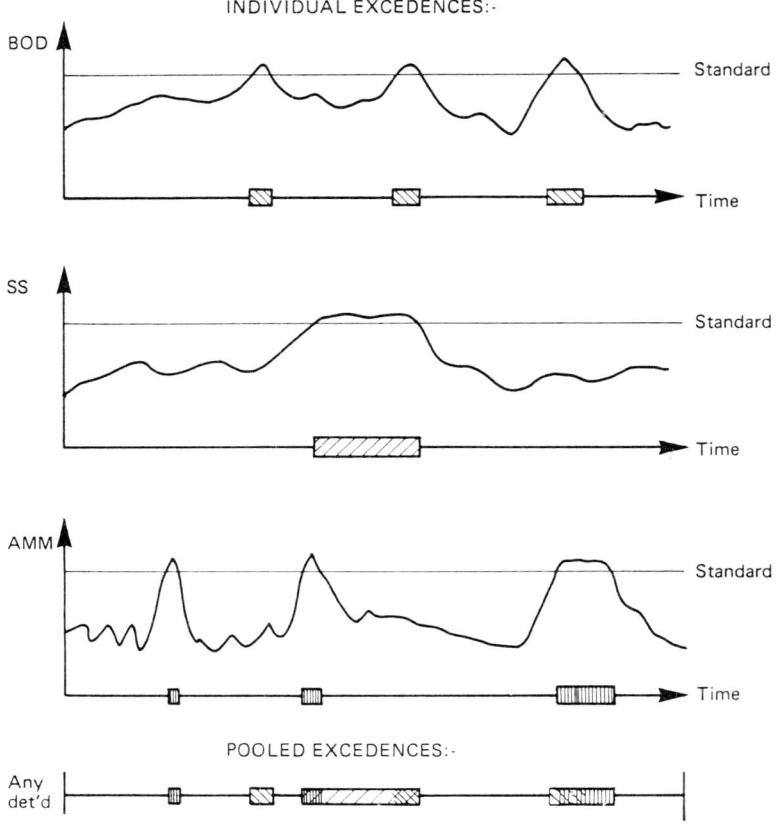

Fig. 8.7 — Multi-determinand compliance.

recent legislation, particularly for heavy metals, to involve ever lower concentrations, the importance of this aspect of compliance is growing. This subject also is addressed in a WRC report [2].

8.3 POLICING

Having explored some of the complexities of the real world, one might be tempted to resort to pure pragmatism, but that would be defeatist and (more to the point) would lay the operating agencies open to attack. A more systematic approach is called for. The 'scientific' approach would be to carry out an assessment of the degree of security afforded by a given sampling investment. It would then be up to the operating agencies, aided by their watchdog groups, to judge the degree of security which is acceptable when balancing risk and cost. Similar decisions have to be made by individuals when deciding on the degree of burglar-proofing of their houses, and by the community and/or Government in allocating resources to police work.

Real life involves a mixture of all approaches. In many cases the monitoring requirements, albeit at a minimum level, are laid down in the legislation. It would be reassuring to feel that these had been derived on a rational basis as discussed above, but the evidence does not support this. In the approaches to domestic pollution control, history plays a large part and international regulations carry a heavy dressing of compromises.

8.3.1 A statistical approach

The application of statistical techniques will not absolve authority from the need to decide an affordable cover, but it *can* quantify the probability of drawing the wrong conclusion (whether pass or fail) from a limited set of samples. There are some inescapable rules which have to be faced about the security bought with a given level of sampling. This is covered in more detail in Appendix A (section 8.6), but one example is discussed here to demonstrate the approach.

There is still considerable debate internationally about the meaning of the term 'MAC', as used in the EC Drinking Water Directive. For the sake of simplicity, it will be assumed that a straightforward 'thou shalt not pass' maximum is intended. The maximum sampling frequency specified is 365 samples per year, that is daily if uniformly spread. It can be shown that if, for a particular determinand, all 365 sample values are below the relevant MAC there is still a 1 in 20 chance that this result could be obtained even when the 'true' compliance of the underlying population is 99.2%. Further, there is a 1 in 100 chance that the true compliance could be as low as 98.7% and still give 365 clear samples. If the sampling frequency is reduced to weekly, the protection is substantially poorer: 100% *sample* compliance can only protect against a *true* compliance of 94% or 91% (depending on whether the confidence level is 1 in 20 or 1 in 100). Figure 8.8 shows this underlying statistical relationship between the desired stringency of assurance and the required number of samples. The circled points identify the statements made earlier in the paragraph.

As Appendix A (section 8.6) shows, similar curves can be constructed for percentile or mean standards. They all make the same point: the narrower the possible gap is to be between observed and true performance, the greater is the required number of samples — and the money to be spent. This point is developed in section 8.3.3.

8.3.2 Application of system knowledge

Monitoring is seldom carried out with a complete absence of knowledge of the system concerned. Sensible application of this knowledge can help in deciding strategy. However, the application of 'gut feeling', or what engineers call

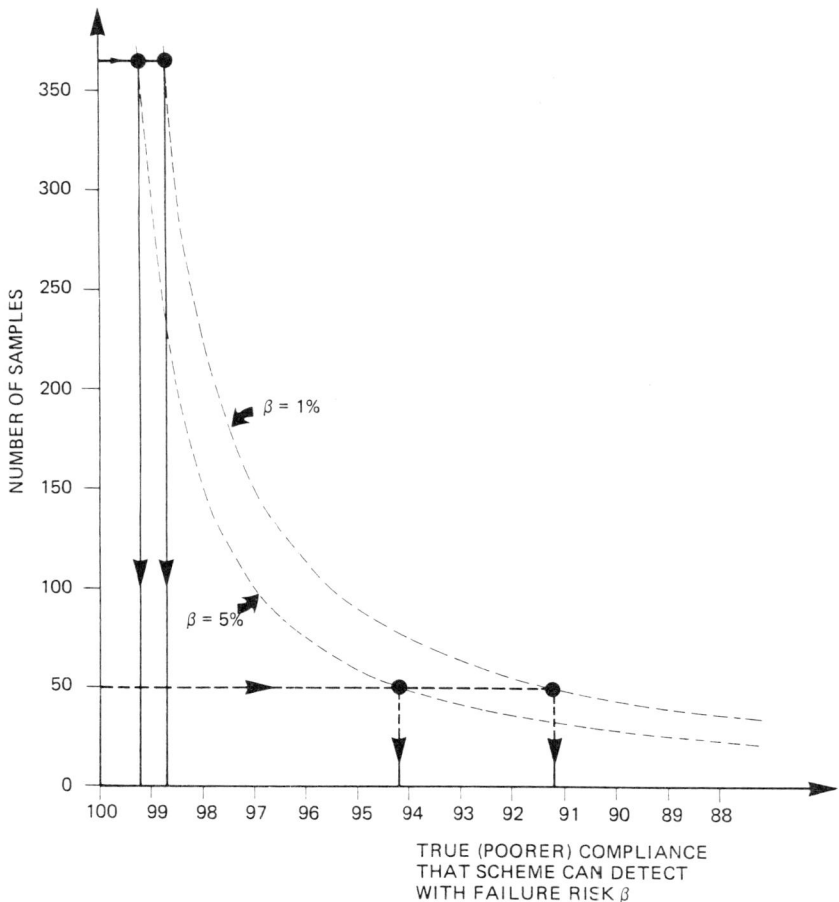

Fig. 8.8 – Relationship between number of samples and degree of security when monitoring against a maximum-type standard.

'engineering judgement', can lead to false assurance. For example, if all the samples give values well below the MAC, as in Fig. 8.9(a), does this imply that the risk of infringement is lower than in Fig. 8.9(b)? The answer is no stronger than 'perhaps' unless something is known about the likely nature of the determinand variations. With a river system or a sewage effluent, where quality is unlikely to vary dramatically, it may be valid to draw additional comfort from Fig. 8.9(a). However, where the quality is less homogeneous — as, for example, with water quality over time at houses in an entire town — it is less certain that there will not be a pocket of houses with a serious problem in spite of samples from other houses showing no problem at all. This point has been more than amply brought out by the national lead sampling exercises.

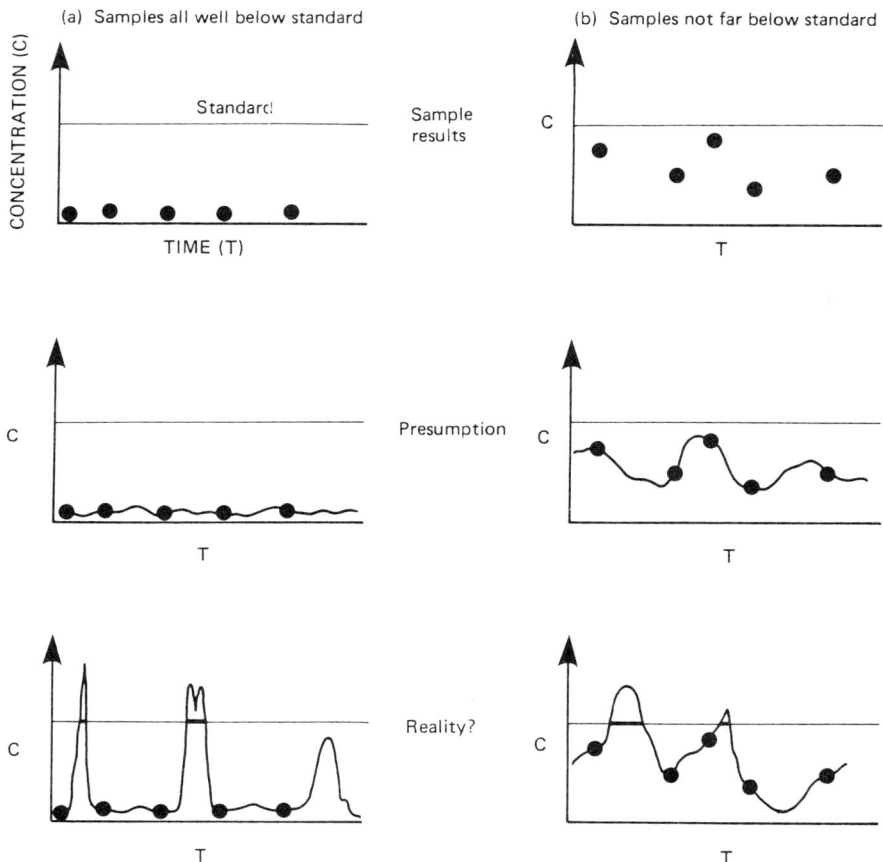

Fig. 8.9 – Inference from 100% sample compliance.

8.3.3 Effect on costs

As previous sections have indicated, more security means more cost. To give some dimension to this, the example of the EC Drinking Water Directive has been followed further. The approximate costs of sampling and analysis have been assessed for a city with a population of 600 000 involving three supply areas fed from different water works. Sampling has been assumed to take place only at the exits from treatment works and not at the tap. Figure 8.10(a) shows the cost of sampling solely for those determinands with obligatory minimum frequencies (C_1 and C_2) laid down in the Directive. Figures 8.10(b), (c) and (d) then show the successive effects of adding other determinands which the Directive leaves to the discretion of the competent national authority, and of sampling for the C_4 determinands at two different frequencies. The derived costs are actually quite small when spread over the population, but so are some of the sampling frequencies.

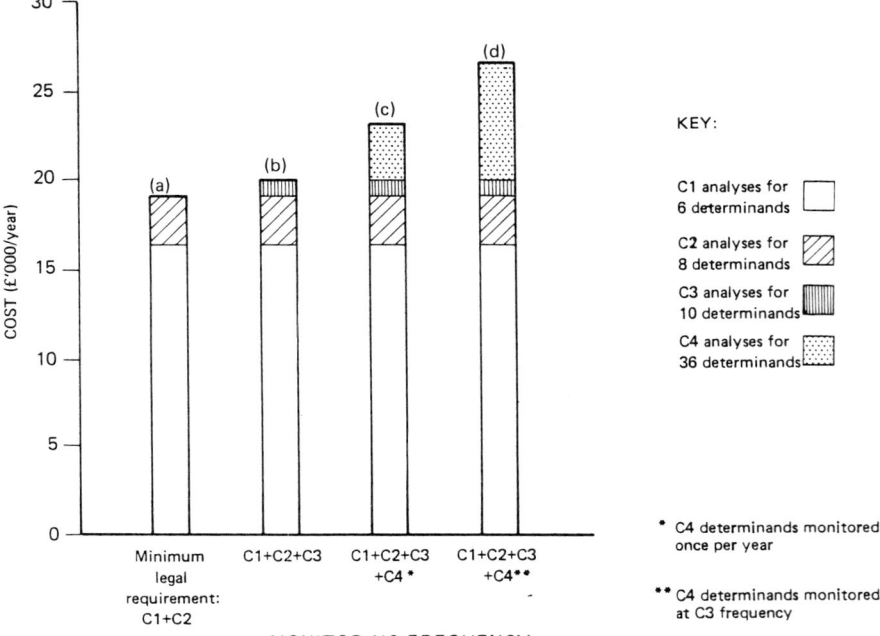

Fig. 8.10 — Costs of obligatory and optional sampling for the Drinking Water Directive (for a population of 600 000).

Extension to tap sampling could escalate these costs substantially. For a single water supply, Fig. 8.11 shows the effects of frequency and number of determinands on annual costs. In the present example, the extreme cost of daily sampling for all determinands represents about 66p per head per annum. The annual cost for the whole of the UK has been estimated on the basis of the minimum sampling required by the EC plus a frequency of once a year for the remaining determinands: this is of the order of £2.4 million. In considering these costs it must be remembered that this is only one of many Directives, and samples for these purposes may in many cases be additional to those called for in normal operations.

8.4 CONCLUSIONS

- The public, environmental pressure groups and operating agencies must all come to terms with the limitations of practical monitoring schemes. At the

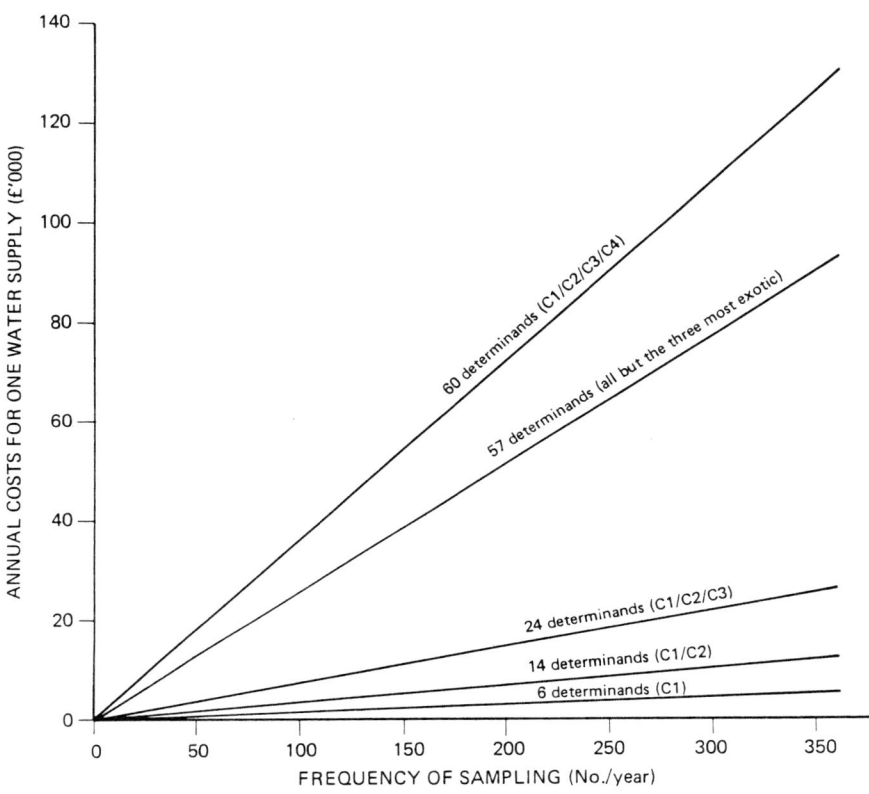

Fig. 8.11 — Effect of sampling frequency and determinand number on annual costs for one water supply.

same time there must be confidence that those entrusted with operational functions will give priority to protection of life in all its forms but will be allowed flexibility in the way they react to standard infringements.
- Fuller use should be made of the scientific knowledge about systems and of the underlying statistical relationships involved in monitoring. In view of the constraints on resources, sampling programmes which look for substances which are unlikely to be present, or which ignore information from previous surveys, should not be encouraged.
- Concentrated surveys would help initially to define the extent and nature of underlying trends, following which a longer-term less intensive sampling programme could then more effectively be developed.
- A uniformly imposed approach which ignores local factors is unlikely to be cost-effective.

8.5 REFERENCES

[1] Ellis, J. C. (1983) Multi-determinand Compliance. Water Research Centre Report 579–M.
[2] Hunt, D. T. E. (1982) Analytical implications of Environmental Quality Standards. Water Research Centre Report 380–M.
[3] Ellis, J. C. (1983) The EQC approach to Water Quality Monitoring: some statistical considerations. Water Research Centre Report 580–M.

8.6 APPENDIX A – STATISTICAL DETAILS

1. Features common to all statistically-based pass/fail schemes

In general, the inevitable presence of random sampling error renders any pass/fail monitoring scheme liable to two types of false conclusion:

(i) The 'Type I' error (α) of judging that the determinand has failed to comply when a *continuous* trace of quality over the reporting period would have been just satisfactory; and
(ii) The 'Type II' error (β) that the determinand is judged to comply when true quality has in fact deteriorated by a stated amount D beyond a just-satisfactory level.

A sampling programme can therefore be defined by three quantities: the Types I and II errors α and β, and the deterioration in quality, D, that it is important to detect. The smaller these are, the more discriminating is the programme – but at a cost in increased sampling effort. The relationship between the prescribed stringency and the consequent number of samples is outlined below for three types of standard: mean, percentile, and maximum. For brevity it is assumed in what follows that the monitoring gives rise to independent, random samples, and that the complications of analytical error (whether bias or random) can be ignored. In practice these simplifying assumptions are often unrealistic, in which event a more general treatment of sample programme design is called for [3].

2. Standard as the mean

When compliance relates to mean quality, the deterioration D that the monitoring must be capable of detecting is most conveniently expressed in multiples of σ, the overall standard deviation of the determinand. Figure 8.A1 plots the required number of samples against D for $\alpha = \beta = 5\%$ and for $\alpha = 10\%, \beta = 1\%$. It will be noticed that the latter prescription is the more expensive in sampling effort. For example, it calls for 52 samples to detect an increase of half a standard deviation in mean quality, compared with 43 samples when the Types I and II errors are 5%.

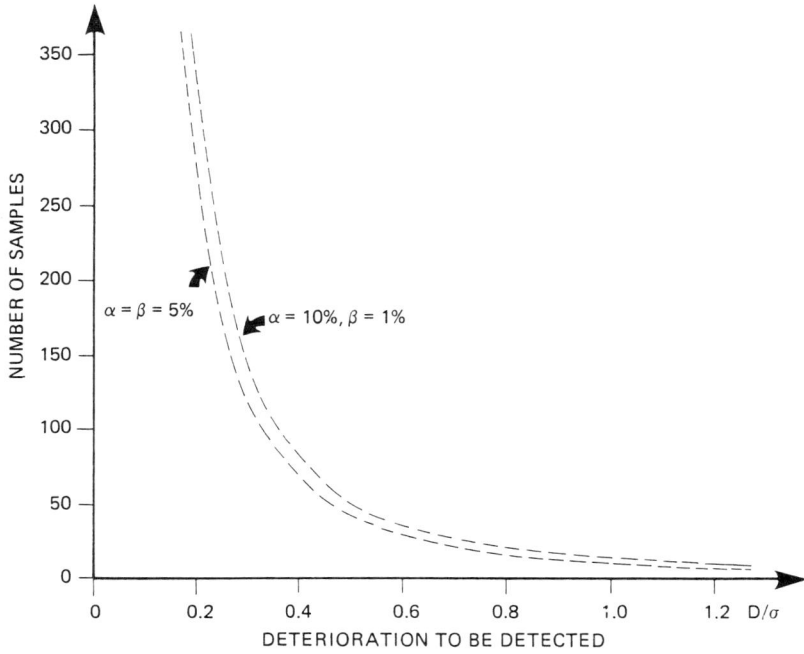

Fig. 8.A1 — Required sample sizes for monitoring the mean.

3. Standard as a percentile

In many specific applications it can be valid to assume that quality follows a particular type of probability distribution (the Normal or the log-Normal, for example); and this leads to more efficient use of sampling effort. Such an assumption would however be inappropriate in the more general context of this paper, and so a non-parametric approach must be taken. This requires the detectable deterioration D to be couched in terms of percentage compliance with the standard. For example, suppose the standard is to be interpreted as a 90 percentile — that is, quality should not stray above the standard for more than 10% of the time. If in fact quality were to exceed the standard for 30% of the time, the true compliance would drop to 70%; and Fig. 8.A2 shows that about 50 samples would be needed to protect against such a deterioration (with α and β risks of 5%).

A similar plot is shown in Fig. 8.A3 for the familiar case of the 95 percentile standard. Now the detection of a three-fold deterioration (that is, from 95% to 85% compliance) requires over 100 samples! This illustrates the general point that higher-percentile standards (if they are to be taken seriously) bring with them more demanding sampling requirements. Putting this another way, suppose

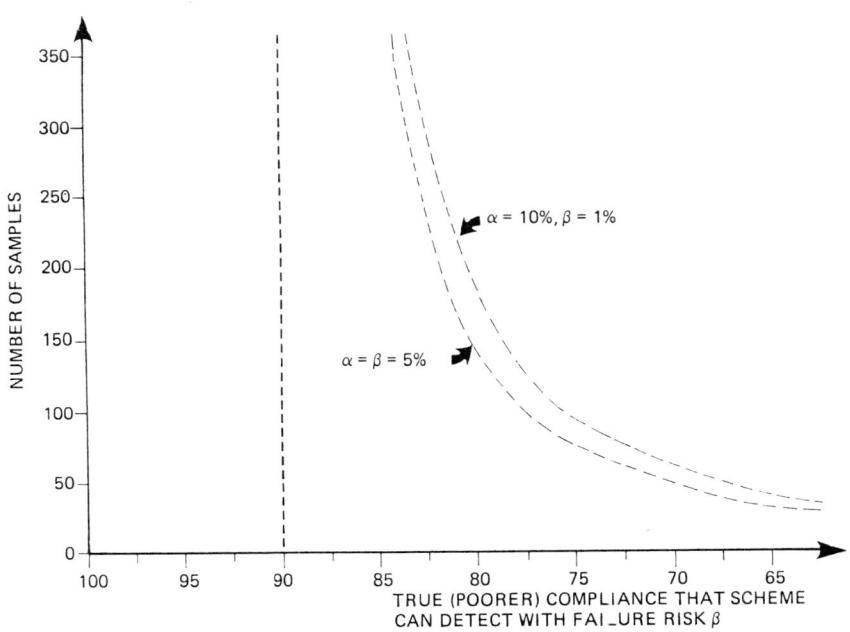

Fig. 8.A2 — Required sample sizes for monitoring the 90 percentile.

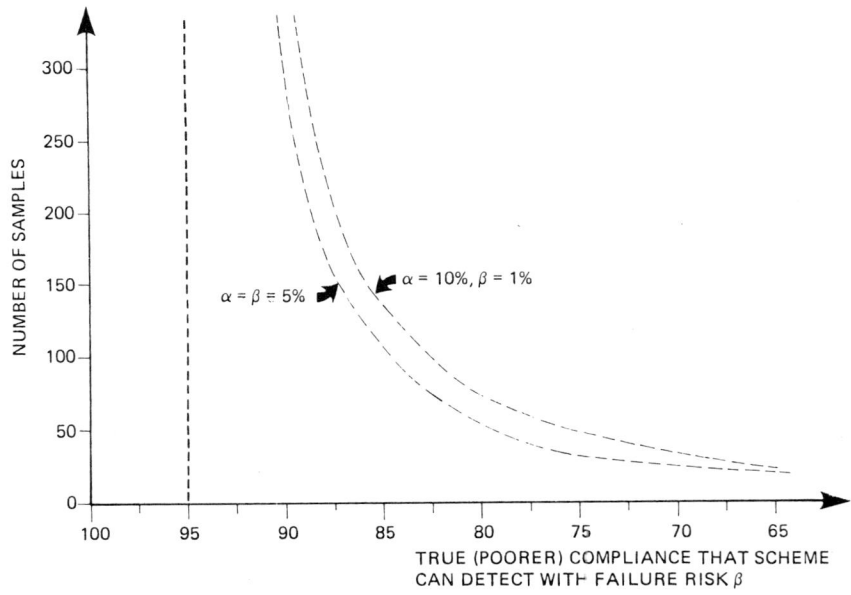

Fig. 8.A3 — Required sample sizes for monitoring the 95 percentile.

a fixed allocation of 50 samples is available. With a *90* percentile standard, this (as remarked earlier) can protect against a *three-fold* deterioration in compliance; but with a *95* percentile standard, 50 samples can guard only against a *four-fold* deterioration (see Fig. 8.A3). In certain circumstances the poorer discrimination probably intended by the choice of a 95 percentile in preference to a 90 percentile standard.

4. Standard as the maximum

When the standard is a maximum concentration never to be exceeded, it can equivalently be regarded as a 100 percentile. Thus the non-parametric method outlined in the previous section for 90 and 95 percentile standards can be applied here too. The only slight difference is that now the Type I error is automatically zero: it is impossible to sound a false alarm because even one solitary failure, out of no matter how many, must be certain proof of non-compliance. In all other respects, Fig. 8.8 in the main text is identical in form to Figs. 8.A2 and 8.A3, and is intepreted in a similar manner. This shows that there is no fundamental *statistical* gulf between a percentile- and a maximum-type standard — a point not always fully appreciated by the legislators.

Discussion

A request was made for more information on where to sample for maximum information and at what frequency to sample to produce data sufficiently accurate to monitor compliance with several determinands. **Mr Ellis** referred to Fig. 8.4 of his chapter and pointed out that the largest histogram related to a sampling point just a few yards from that regional laboratory. Approximately twice as many samples were taken from that sampling point than from any other. Analysis of the data showed that this site was one of a batch of six sites along the river where no significant differences could be found. This indicates that although there may be other good reasons for sampling at these points, there is no statistical need for all six sites to be used, and that one or two sites could be left out of the sampling programme with little loss of information. Perhaps it would then be reasonable to take a larger number of samples at the remaining sites.

The question of sampling frequency is very difficult to answer because it depends on the choice of precision and confidence required — for example $\pm 5\%$ with 95% confidence, or $\pm 20\%$ with 80% confidence — and that choice is to some extent arbitrary. For this reason WRC have been approaching the problem from the opposite direction by asking what levels of precision and confidence can actually be achieved from existing sets of historical data. Often the answers from such retrospective analyses point very clearly to practical ways in which future sampling can be made more effective.

The question was raised as to whether bacteriological sampling programmes should be carried out in the same way as chemical sampling studies which, traditionally, use a thin blanket-cover approach rather than an intensive study at a few sites. **Dr Miller** said that every case should be looked at with regard to the determinand being measured and the objectives of the study. By and large the water authorities do not have the resources to sample all the determinands at every place. However, the microbiological sampling requirements might be considered a special case here. The authorities feel obliged to analyse at least once every determinand so as to be able to respond to the public's demand for information. However understandable this approach is, does not really advance our knowledge. There is a need for case studies and concentrated examinations to define the problem.

The problem was raised of occasional very bad samples which, though indicative of great environmental damage, would not prevent the discharger's effluent from meeting the required percentage compliance if they were associated with a sufficiently large number of acceptable samples. **Mr Ellis** agreed that this was an inevitable weakness of the percentile type of approach, and, recalling a slide shown in Dr Miller's presentation, observed that furthermore the non-compliant samples could all occur in one consecutive burst.

There was discussion about the uncertainties around safe or acceptable levels for some chemicals, and the economic consequence of actions such as taking waters out of supply based on these uncertainties. Was there a need for bringing together at an early stage the scientific facts upon which limits are based and the economic consequences of any proposed limits? If so, how could this be done?

Dr Packham replied that there is a demand for limits; people need some guidance and reassurance. A lot of standards are produced in an attempt to answer that demand. However, there is a tendency for a single number to become enshrined and legalised. That is the crux of the problem. The Guidelines for Drinking Water Quality (WHO) will be a three-volume document, the second (and largest) volume being dedicated to explaining in detail the basis on which limits are set. Dr Packham agreed that the limits require linking to some cost–benefit or risk analyses in order to help people make the decision they have to make.

Sir Frederick Warner recommended the British Standards Institution Handbook No. 22, which is a collection of all the standards related to quality assurance. The methodologies of the standards, data collection and handling are all given in detail.

CHAPTER 9

The role and activities of the Oslo and Paris Commissions

F. BJERRE and **P. A. HAYWARD**
Oslo and Paris Commissions

9.1 INTRODUCTION

The subject of this Conference is the protection of the aquatic environment. This may be divided for the purpose of discussion into at least three 'geographical' compartments: inland waters, groundwater, and marine waters. It is for the protection of the marine environment that international control is not only justified but is indeed the only appropriate level at which action can be initiated. The recent UN Conference on the Law of the Sea reaffirmed the principle that the sea is the common heritage of mankind and part of the justification for the lengthy negotiations was the need to ensure the preservation of that heritage for the benefit of all. It is for the protection from pollution of part of that 'common heritage' – the North East Atlantic region – that the Oslo and Paris Commissions were established by international treaty.

Much of the discussion during this Conference will focus upon the standards imposed by governments on the water industry to control the input of polluting substances in order to ensure that the aquatic environment is adequately protected. When it is the protection of the marine environment that is under consideration, it is pertinent to ask whether the national measures which have been taken would have been made had it not been for the stimulus given by the international conventions. It is true that some European States can trace the development of their environmental legislation back to the thirteenth century. However, surely a feature of our times is the greater emphasis which has been placed by all the European nations on stronger environmental legislation. In part this is the logical response to an increasingly complex industrial development, particularly in the chemicals field; in part it is the result of greater public environmental consciousness, which is perhaps more prominent in some countries than in others.

Concerning the protection of the North East Atlantic in general and the North Sea in particular, there is strong evidence that the international consensus achieved in the adoption of the Oslo and Paris Conventions provided the stimulus for the subsequent legislation at national level. A cursory examination of the dates of national legislation for the protection of the marine environment among the Parties to the Oslo and Paris Conventions indicates that the national legislation which has been developed arose from the need to give legislative effect to the international commitments and obligations to which the States had voluntarily bound themselves by ratifying these conventions.

A brief look at the history of the negotiations which led to the Oslo Convention will serve to illustrate the need for protection and control to be exercised at international level. During the late 1960s there were no European States which had legislation to control the dumping of wastes from ships which, at that time, was a cheap and easy method for industrialists to use to dispose of their wastes. Gradually, waste dumping began to be brought under control by voluntary codes of practice but several States felt the need for more stringent regulation — provided that the restrictions applied equally to everyone. Initially there were difficulties in reaching international agreement on how the activity should be controlled, what substances should be prohibited from dumping and what conditions should be imposed. The need for the international negotiations to reach positive conclusions was demonstrated by the case of the *Stella Maris*. This was a Dutch coaster which left the port of Rotterdam with the intention of dumping 650 tons of chlorinated hydrocarbons in the northern part of the North Sea. The proposed dump site gave rise to objections from the Norwegian government and further proposed sites elicited objections from the governments of Iceland and Ireland. Eventually the coaster returned to harbour with its cargo intact. The *Stella Maris* incident served to demonstrate first, that the practice of ocean dumping needed to be legally regulated and, second, that national measures were not enough and that it was essential for the control to be agreed and exercised at international level.

9.2 THE OSLO CONVENTION

The national viewpoints expressed during the negotiations on the Oslo Convention in 1971 and 1972 demonstrated the difference of philosophy which existed then among the States of Europe and which is still, to a large extent, prevalent today. Some States held the view that the dumping of wastes at sea should be totally prohibited, with the possible exception of dredged spoils, large quantities of which need to be disposed of in order to maintain easy access to ports and harbours. Other countries were of the opinion that the resources of the sea may legitimately be used for the disposal of wastes provided that any adverse environmental effects are kept within acceptable limits. Despite these differing viewpoints on whether or not dumping should be allowed, all the States which ratified the Oslo Convention agree upon the necessity of preventing 'pollution' by dumping.

In Article 1 of the Oslo Convention

'the Contracting Parties pledge themselves to take all possible steps to prevent the *pollution* of the sea by substances that are liable to create hazards to human health, to harm living resources and marine life, to damage amenities or to interfere with other legitimate uses of the sea'.

This 'definition' of pollution is based on that formulated by GESAMP (Joint Group of Experts on the Scientific Aspects of Marine Pollution) in 1970 and which has been used, in variously modified forms, in most marine pollution conventions negotiated since that date. Essentially the definition of pollution does not prohibit the input of contaminating substances as such; rather it requires there to be evidence of some degree of harm. It is perhaps worth noting that this GESAMP definition of pollution was first formulated by scientists and that it has since been adopted by the administrators and legislators. The reason for its continued usefulness is that it provides helpful guidelines on which to base subsequent action and policy without being too rigid.

Although the definition of pollution may be open to interpretation, the Oslo Convention does lay down a number of ground rules which did not exist before. Perhaps the most important principle deriving from the Oslo convention, which is to be found in practically all subsequent international treaties and directives on marine pollution, is the notion that substances can be classified into 'black' and 'grey' lists. Annex I to the Convention therefore lists those substances — the 'black' list — the dumping of which is prohibited. The prohibited substances are organohalogen and organosilicon compounds and compounds which may form such substances in the marine environment, mercury and cadmium and their compounds, substances which have been agreed between the Contracting Parties as likely to be carcinogenic under the conditions of disposal (as yet none such have been agreed) and persistent plastics and other persistent synthetic materials which may float or remain in suspension in the sea.

The drafters of the Convention did acknowledge that there needed to be some exemption to the rigid application of this list. Mercury and cadmium are naturally occurring elements and would be present in some degree in practically all wastes dumped at sea. Therefore the Convention provides that the prohibition of the dumping of substances listed in Annex I shall not apply where these substances occur as 'trace contaminants' in waste to which they have not been added for the purpose of being dumped. Furthermore, in respect of organohalogen and organosilicon compounds, the prohibition exempts those compounds 'which are non-toxic, or which are rapidly converted in the sea into substances which are biologically harmless'.

Such exemptions could of course give rise to abuse. At its First Meeting in 1974, the Oslo Commission therefore established a Prior Consultation Procedure (PCP) whereby Contracting Parties proposing to issue a permit for the dumping of wastes containing Annex I substances, on the assumption that they qualified for the exemptions referred to above, must first inform the other members of the Commission. In 1976 the Commission adopted a set of test procedures to evaluate the toxicity, biodegradability and bioaccumulation potential of the Annex I fraction of wastes which Contracting Parties were required to apply in order to demonstrate the 'harmlessness' of the proposed dumping operation. Revised test procedures, modified in the light of more recent scientific developments, were adopted by the Commission in 1982 and 1983.

The Commission has also modified the Prior Consultation Procedure itself in the light of experience. In particular, Contracting Parties are now required to give documented information as to the alternative disposal routes to marine dumping that have been considered and rejected by the appropriate authorities. The revised PCP also defines certain numerical values for inorganic mercury and cadmium. At the lowest level, the values define what amounts to a definition of 'trace contaminant' for inorganic mercury and inorganic cadmium, below which dumping may be permitted without recourse to the Prior Consultation Procedure. At the higher level, the Contracting Party concerned must follow the Prior Consultation Procedure if it considers that the dumping of wastes containing such amounts of inorganic mercury and cadmium ought nevertheless to be permitted. Dumping permits issued for wastes containing inorganic mercury and cadmium between these two limits must be notified to the Secretariat of the Oslo Commission without delay for the information of the other Contracting Parties.

The second fundamental change which the Oslo Convention introduced was the requirement that all dumping operations shall require the approval of the appropriate national authority or authorities. In the case of wastes containing 'significant quantities' of Annex II substances (subsequently agreed by the Commission as representing more than 0.1% of the waste for disposal), the Convention requires a so-called 'specific permit'. Annex II lists those substances and materials which may be dumped but which require 'special care'. They include arsenic, lead, copper, zinc and their compounds, cyanides and fluorides, and pesticides and their by-products not covered by Annex I. It also includes containers, scrap metal and tar-like substances likely to sink to the sea bottom and other bulky wastes which may present a serious obstacle to fishing or navigation. Annex II also includes substances which, though of a non-toxic nature, may become harmful due to the quantities in which they are dumped, or which are liable to seriously reduce amenities. The dumping of bulky wastes must always be carried out in deep water, that is not less than 2000 metres and at least 150 nautical miles from the nearest land.

The third significant change introduced by the Convention is Annex III, which specifies certain provisions which must be applied by Contracting Parties in issuing permits and approvals for dumping. Annex III identifies a number of characteristics of the waste which must be taken into account before a permit is issued; it also specifies certain characteristics of the dumping site and method of deposit which need to be borne in mind. In the general considerations and conditions governing the issue of permits, it specifies that the practical availability of alternative means of disposal or elimination must be taken into consideration.

The Convention also establishes a Commission made up of representatives of each of the Contracting Parties. The duties of the Commission are enumerated in Article 17:

(a) to exercise overall supervision over the implementation of the Convention;
(b) to receive and consider the records of permits and approvals issued and of dumping which has taken place, and to define the standard procedures to be adopted for this purpose;
(c) to review generally the condition of the seas within the area to which the Convention applies, the efficiency of the control measures being adopted, and the need for any additional or different measures;
(d) to keep under review the contents of the Annexes to the Convention, and to recommend such amendments, additions or deletions as may be agreed;
(e) to discharge such other functions as may be appropriate under the terms of the Convention.

At its First Meeting in 1974 the Commission adopted the first standard forms of report for providing information on dumping permits. Since then the procedures have been modified in the light of experience and the Commission now receives reliable data from the Contracting Parties on the dumping activities carried out. At its meeting in 1983 the Commission examined a report by the Secretariat on the dumpings carried out in the six years from 1976 to 1981. Although the data for the early years were incomplete (because in some countries the licensing procedures to implement the Oslo Convention were not yet in force), the Commission was able to conclude that

'from the figures available it appeared that there was some evidence for a tendency in the amounts of industrial wastes being dumped to decrease and that a further reduction might be expected in future years. The fluctuation in the amounts of dredged materials dumped was caused by natural circumstances. The amounts of sewage sludge being dumped were more or less constant. The Commission agreed upon the importance of reducing the concentration of harmful substances in dredged materials and sewage sludge. The Commission agreed that its activities in the future would be directed at achieving a greater reduction in the pollution of the marine environment from dumping and incineration activities. The many instruments now available to the Commission and to the Contracting Parties, together with their firm intentions, provided the favourable background necessary for the realization of the new efforts'.

Mention has just been made of the practice of incineration at sea. This is a method of disposing of organochlorine wastes which was developed during the 1970s. In 1977 the Oslo Commission adopted the first code of practice governing incineration at sea. Subsequently, legally binding regulations were adopted by the Contracting Parties to the London Dumping Convention (the instrument governing dumping activities at the global level) and in 1981 the Oslo Commission agreed upon its own legally binding rules on incineration at sea. The protocol to amend the Convention and a new Annex IV containing the incineration rules, were opened for signature in Oslo in March 1983.

The purpose of the amendment to the Convention is essentially threefold. First, it amends the definition of 'dumping' to make it absolutely clear that this includes any deliberate disposal into the sea from ships and aircraft, including through the medium of incineration at sea. Second, in order to contain the practice of incineration at sea, it limits the substances and materials which may be incinerated to organohalogen compounds, and pesticides and their by-products other than organohalogen compounds. Other substances may be the subject of an incineration permit but in such cases the Contracting Party proposing to issue a permit for incineration would first have to follow a Prior Consultation Procedure with the other members of the Commission and justify the need for incineration at sea. Third, the new Annex to the Convention stipulates that the Commission shall meet before 1 January 1990 to establish a final date for the termination of incineration at sea.

The Oslo Convention's rules governing incineration are therefore more restrictive than the comparable legal instrument adopted within the London Dumping Convention. This is due to the strongly held views of some Contracting Parties that the practice of incineration at sea is only acceptable as a temporary means of waste disposal and that it should be ended as quickly as possible. One of the chief reasons for this concern is that it is more difficult to exercise control over incineration operations at sea than on land. Secondly, because of the impossibility of installing efficient 'scrubbing' systems to the stacks of incineration vessels, the destruction efficiency of land-based incinerators is generally greater than can be achieved by the incineration vessels.

9.3 IS THE OSLO CONVENTION WORKING?

The enforcement provisions of the Convention are contained in Article 15. Among other things, this Article requires each Contracting Party to ensure compliance with the provisions of the Convention:

(a) by ships and aircraft registered in its territory;
(b) by ships and aircraft loading in its territory the substances and materials which are to be dumped;
(c) by ships and aircraft believed to be engaged in dumping within its territorial sea.

The enforcement provisions of the Convention thus focus, quite logically, on the control of the dumping operation itself. It is rare for dumping operations to be carried out by one Contracting Party within the territorial sea of another. Control of dumping and incineration operations by the flag State and by the port State, particularly the latter, can be readily enforced. It is evident from the statistics available to the Commission that such control is on the whole being exercised by the Contracting Parties to the Convention.

It is evident, however, that in the process of dumping and incineration at sea there is one party which is omitted from the Convention's provisions, viz. the waste producer and the country of production. In order to establish the extent to which this was a problem, and particularly the extent to which wastes were 'imported' from States which were not Parties to the Oslo Convention, the Commission established a Working Group. This group concluded that the transfrontier movement of wastes for sea disposal mainly concerned wastes destined for incineration at sea rather than dumping. Indeed, 70% of the wastes incinerated at sea are exported to the country of loading, mainly Belgium. There is less evidence of wastes specifically exported and imported for dumping and such transfrontier movement of wastes as exists is effectively controlled by the two countries chiefly concerned, the Federal Republic of Germany and the Netherlands, although the transit of wastes as such is not at present specifically controlled by legislation in any Contracting Party. One important conclusion reached by the Working Group and confirmed by the Commission's statistics is that only about 5% of wastes destined for incineration at sea originate from non-Contracting Parties.

The Commission has therefore concluded that no action is justified at the present time to amend the Oslo Convention, which can only be concerned with the export/import of wastes destined for dumping or incineration at sea. The Commission will review the position again in 1985 but, in the meanwhile, the wider problem of the transfrontier movement of wastes is being discussed within the framework of the EEC.

Is the Oslo Convention working effectively? If this question were to be asked of any of the environmental groups which are concerned with the protection of the marine environment their answer would probably be no. Even these groups would probably admit that within the limits of the Convention text itself, however, the answer would be in the affirmative. Dumping operations are now regulated by permit and there is no abuse of the exemptions for the dumping of Annex I substances. However, within the wider context, the dumping of acid wastes from the titanium dioxide industry is still permitted and this is the subject of some concern, not only to the environmental groups, but also to the Contracting Parties to the Convention. This leads us to the examination of how well another provision of the Convention has been implemented, namely scientific and technical research.

Under Article 12 of the Convention, 'the Contracting Parties agree to establish complementary or joint programmes of scientific and technical research, including research on alternative methods of disposal of harmful substances, and to transmit to each other the information so obtained'. At its First Meeting the Commission established a Standing Advisory Committee for Scientific Advice (SACSA) with the responsibility, among others, of keeping under review the advance of scientific and technical knowledge in so far as this may assist the work of the Commission. As yet there has been little progress in the development of complementary or joint programmes of research, although SACSA has provided the forum for an exchange of information on alternative methods of disposal to the dumping of substances such as fly ash, sewage sludge and dredged spoils. As regards the dumping of TiO_2 wastes, SACSA has been informed to research proposals in the Federal Republic of Germany designed to reduce the dumping both of the dilute acids (diluted sulphuric acids containing impurities) and the so-called green salts (ferrous sulphate). But the wastes from the titanium dioxide industry are not only dumped at sea; in some countries they are discharged from pipelines and this concerns the work of the Paris Commission.

9.4 THE PARIS CONVENTION

The preamble to the Oslo Convention noted that marine pollution has many sources, including discharges through rivers, estuaries, outfalls and pipelines within national jurisdiction. Thus it was that the States which elaborated the Oslo Convention went on to tackle the more difficult problem of preventing pollution from land-based sources. The Convention was opened for signature in 1974 and was signed by all those States which have since ratified the Oslo Convention with the exception of Finland (which has no land-based discharges to the Paris Convention area) but with the addition of the EEC as a Contracting Party. Indeed, the preamble to the Council Directive of 4 May 1976 on pollution caused by certain dangerous substances discharged into the aquatic environment of the Community (76/464/EEC) cites as part of its *raison d'etre* the need to ensure the co-ordinated implementation of the Paris Convention and other draft conventions designed to protect international watercourses and the marine environment from pollution. The preamble to the Paris Convention itself justifies the action being taken to control marine pollution from land-based sources 'as part of progressive and coherent measures to protect the marine environment from pollution, whatever its origin'.

The Paris Convention applies to broadly to the same maritime area as the Oslo Convention. It covers the high seas, the territorial seas of the Contracting Parties, and waters on the landwork side of the baselines from which the breadth of the territorial sea is measured up to the fresh water limit in the case of watercourses.

The Convention defines 'pollution from land-based sources' as the pollution of the maritime area:

- through watercourses;
- from the coast, including introduction through underwater or other pipelines;
- from man-made structures placed under the jurisdiction of a Contracting Party within the limits of the area to which the present Convention applies.

Although this is not specifically stated, the implication is that the pathway of pollutants from land-based sources will be through the water medium. It is noteworthy that no specific mention of atmospheric pollution is to be found in the Paris Convention text and it seems reasonable to conclude that the drafters of the Convention did not consider this to be a major source of pollution of the marine environment — or at least a source which could be readily controlled — when the Convention was drafted. The omission of any specific reference to controlling pollution from or through the atmosphere has been the subject of some discussion in the Paris Commission in recent years as more and more evidence has become available of the pollution load from the atmosphere. The only clause in the present Convention text which would appear to lend support to the Paris Commission taking action in this field is the reference in Article 1 which defines pollution of the sea as 'the introduction by man, directly or *indirectly*, of substances or energy . . . .'

The Commission has, however, decided to establish a Working Group to collect and evaluate national and international information on the airborne input of pollutants to Convention waters. It will be the task of this Working Group to identify the possible technical means, and their approximate costs, by which atmospheric emissions might be controlled. The Working Group will also assess the information on pollution of the Paris Convention area from or through the atmosphere, in comparison with other pathways of pollutants from land-based sources. It will also identify research gaps and recommend to the Commission what action would be required in order to meet the objectives of the Convention. The Commission recognises that if these studies suggest it would be useful and practical to control airborne pollution of Convention waters, then it will be necessary for the Commission to consider whether changes are needed to the Convention text in order to take specific measures in this field.

As with the Oslo Convention, the Paris Convention works essentially by considering substances according to their characteristics for persistence in the marine environment, toxicity or other noxious properties and tendency to bioaccumulation. List I, the 'black list', includes those substances which are not readily degradable or rendered harmless by natural processes and which either give rise to dangerous accumulation of harmful material in the food chain, or endanger the welfare of living organisms causing undesirable changes in the marine ecosystems, or interfere seriously with the harvesting of sea foods or with other legitimate uses of the sea. They were also listed because the drafters of the Paris Convention considered that pollution by these substances necessitated urgent action. The substances included in the black list are organohalogen compounds; mercury and mercury compounds; cadmium and cadmium compounds; persistent synthetic floating materials; and persistent oils and hydrocarbons.

The substances listed in Part II of Annex A of the Convention, the 'grey list' are included because they exhibit similar characteristics to the substances in Part I and require strict control, but nevertheless seem less noxious or are more readily rendered harmless by natural process. The substances listed include the organic compounds of phosphorus, silicon and tin; elemental phosphorus; non-persistent oils and hydrocarbons; the following elements and their compounds: arsenic, chromium, copper, lead, nickel and zinc; and organoleptic substances (substances likely to have a deleterious effect on the taste and/or smell of sea foods).

There is a third category of substances included in the Paris Convention: radio-active substances, including radio-active wastes. Although the Convention text recognises that such substances should be subjected to stringent controls and that pollution from them should be prevented or eliminated, the Convention recognises that they are already the subject of research, recommendations and, in some cases, measures under the auspices of several international organisations and agencies. The Paris Commission is currently reviewing the extent to which discharges of radio-active substances are indeed covered by measures taken within the framework of other international fora and the extent to which monitoring within the vicinity of the discharge points is carried out.

The crux of the Paris Convention is contained in Article 4 which requires the Contracting Parties to implement, jointly or individually as appropriate, programmes and measures for the elimination, as a matter of urgency, of pollution by the substances listed in Part I of Annex A and for the reduction or, as appropriate, elimination of pollution by substances listed in Part II of Annex A.

It should be noted that the Paris Convention does not prohibit the discharge of a blacklisted substance as such. What is required by the Convention is the elimination of *pollution* by blacklisted substances. This is an important distinction from the Oslo Convention which positively prohibits the dumping of blacklisted substances apart from the exceptions already referred to. An important requirement of Article 4, however, is that listed substances shall only be discharged after approval has been granted by the appropriate authorities within each Contracting State.

The scope for undertaking 'programmes and measures' as outlined in Article 4 is broad. The Convention text identifies four ways in which action can be taken

- specific regulations or standards governing the quality of the environment;
- discharges into the maritime area and such discharges into watercourses as affect the maritime area;

- the composition of substances and products;
- the use of substances and products.

Article 4 requires Contracting Parties to take the latest technical developments into account when defining the programmes and measures to be adopted. It also stipulates that the reduction and elimination programmes shall contain time-limits for their completion.

As with the Oslo Convention, the Paris Convention establishes a Commission composed of representatives of each of the Contracting Parties. The duties of the Commission, as prescribed in Article 16 of the Convention, are:

(a) to exercise overall supervision over the implementation of the Convention;

(b) to review generally the condition of the seas within the area to which the Convention applies, the effectiveness of the control measures being adopted and the need for any additional or different measures;

(c) to fix, if necessary, on the proposal of the Contracting Party of Parties bordering on the same watercourses and following a standard procedure, the limit to which the maritime area shall extend in that watercourse;

(d) to draw up, in accordance with Article 4 of the Convention, programmes and measures for the elimination or reduction of pollution from land-based sources;

(e) when pollution from land-based sources originating from the territory of a Contracting Party by substances not listed in Part I of Annex A of the Convention is likely to prejudice the interest of one or more of the other Parties, to consider the question if requested by any Contracting Party concerned and to make recommendations with a view to reaching a satisfactory solution;

(f) to receive and review information on monitoring programmes, on measures taken to ensures compliance with the Convention and reports on the substances listed in the Annexes which are liable to find their way into the maritime area;

(g) to make recommendations regarding any amendments to the lists of substances included in Annex A to the Convention;

(h) to discharge such other functions, as may be appropriate, under the terms of the Convention.

The more dynamic role of the Paris Commission is evident from this list of functions which it is obliged to perform. The role of the Oslo Commission is essentially to regulate at international level an existing activity; there is little incentive in the Oslo Convention text requiring the Commission to take an innovative role although this has in fact happened, for example as with regard to controlling incineration at sea. However, the Paris Convention text itself requires the Commission to take more initiatives and to play a legislative role, most notably in the injunction to draw up programmes and measures for the elimination or reduction of pollution from land-based sources. As is clearly laid down in Article 18, the programmes and measures adopted by the Paris Commission have legal effect and, unless otherwise specified by the Commission, shall be applied by all Contracting Parties 200 days after their adoption.

9.5 HOW EFFECTIVE IS THE PARIS COMMISSION?

The Paris Convention entered into force in 1978 but its activities commenced immediately following the signature of the Convention in 1974. It is not possible to list all the achievements of the Commission since that time but in any attempt to evaluate the usefulness of the Commission it is relevant to look at the extent to which the provisions of Article 4 are being implemented.

As has already been noted, Article 4 provides for the elimination of pollution by both environmental quality standards and discharge standards. Although sanction to this dual approach was contained in the Convention text itself, its first practical application came with respect to controlling the discharges of mercury from the chloralkali industry. The Commission agreed at its First Meeting to fix uniform emission standards for mercury discharges from both existing and new chloralkali plants. It also provided that, as a derogation from the application of the uniform emission standards, a Contracting Party can proceed by the environmental quality objectives approach, provided that it can demonstrate to the Paris Commission that the environmental standards agreed by the Commission are being met and continuously maintained throughout the area which might be affected by the discharges. The Commission agreed that proposals for quality objectives should be formulated for organisms, 'as soon as possible for water and, if appropriate, for sediments'.

In dealing with the mercury problem the Paris Commission has adopted a stepwise approach. Rather than attempting to solve all problems in one move the Commission agreed that it was preferable to make balanced progress on both the EQO and UES approaches in a piecemeal fashion rather than await the situation where all aspects of the problem could be covered in one comprehensive package. Thus it was that the Paris Commission was able to reach decisions on an environmental quality standard for organisms and limit values for existing chloralkali plants in advance of the EEC directive on the subject. Of course, those States which are both Contracting Parties to the Paris Convention and Members of the European Communities would not have been able to accept measures within the Paris Commission if they had not been able to accept identical measures within the EEC, and vice versa. Nevertheless, the negotiations on mercury demonstrated certain advantages in States being able to discuss the same problem within two fora, although equally it demonstrated the need to avoid any unnecessary duplication of effort.

Work on the prevention of pollution by cadmium illustrates how this duplication can and has been avoided. Negotiations within the EEC have focussed on the adoption of a directive on cadmium discharges and environmental quality standards for cadmium, under the parent directive 76/464/EEC. The Paris Commission Working Group studying the problem deliberately set such considerations on one side and is concentrating on another of the methods envisaged by Article 4 of the Convention, notably product control-oriented measures. In particular, the Working Group is looking at the use of cadmium in pigments, PVC stabilisers and the surface coating industry.

The possibility of formulating programmes and measures governing the use of substances is another approach which has been used to good effect by the Paris Commission. Polychlorinated biphenyls and polychlorinated terphenyls are persistent products and the Paris Commission established a Working Group to review the various sources of PCBs and PCTs and to investigate in particular the acceptability of substitutes in the various uses which are currently permitted by the EEC legislation of 1976.

Following this review, the Paris Commission has accepted that there are no technical or economic reasons why the use of PCBs in transformers, large and small capacitors, and heat-transmitting fluids should not be prohibited in new equipment within a very short period of time. The Commission will consider at its next meeting the setting of a precise data for implementing the phasing out of these uses.

The Commission has also agreed to recommend that the use of PCBs as a chemical feedstock or as intermediates in synthesis processes should be prohibited by all Contracting Parties. The Commission has recommended that PCTs used in tooling compounds during the manufacture of jet engine turbine blades, semiconductors and optical equipment should also be phased out of use as soon as possible. Furthermore, the Commission has decided that Contracting Parties should be encouraged to embark at once upon plans for phasing out the existing uses of PCBs and PCTs. The Commission will review in 4 years' time progress on the phasing out of both new and existing uses of PCBs and PCTs.

In addition to the important function of drawing up 'programmes and measures', the Commission has also to receive and review the information submitted by the Contracting Parties 'in accordance with a standard procedure' on the substances listed in the Annexes to the Convention and liable to find their way into the maritime area. The Commission has made a start on implementing this requirement of the Convention by requiring the Contracting Parties to report annually, by means by standard report formats, the inputs of hydrocarbons from offshore oil and gas platforms, mercury from the chloralkali industry, and the discharges of wastes from the titanium dioxide industry. In addition, the Commission has decided to survey in depth every three years the discharges from oil refineries, from oily waste reception facilities in ports and from offshore platforms. Efforts are also being made to gather information

Ch. 9]   The Role and Activities of the Oslo and Paris Commissions   155

on the input of cadmium, mercury and PCBs from land-based sources by means of a standard report form but there are serious difficulties in collecting information, notably in agreeing a methodology for calculating the input of such substances from rivers.

9.6 MONITORING THE MARINE ENVIRONMENT

One of the most important obligations placed upon the Contracting Parties to both the Oslo and Paris Conventions is the requirement to establish complementary or joint programmes for monitoring the distribution and effects of pollutants in the area to which the Conventions apply. Recognising their joint needs in this respect, the Comissions decided to establish a Joint Monitoring Group whose main task has been to establish the principles and procedures of the Joint Monitoring Programme (JMP). Under this programme the levels of mercury, cadmium and PCBs in marine organisms and mercury and cadmium in sea water are collected and evaluated for approximately 50 zones within the Convention area. In order to ensure that the results obtained are comparable, the participating laboratories must have performed successfully in international intercalibration exercises.

The annual results of the analyses for the period 1979 to 1981 show that living organisms in certain marine areas contain higher levels of mercury, cadmium and PCBs than would be expected under unpolluted conditions. The levels of cadmium and mercury in sea water are not abnormal for coastal or estuarine waters. Although there are few zones in the JMP where the concentrations observed give cause for concern, the Programme as such provides a useful opportunity for all Contracting Parties to review the state of the marine environment on a comparable basis. The way in which the Contracting Parties have worked together to establish the Joint Monitoring Programme is an example of successful international co-operation.

9.7 CONCLUSIONS

It is now approximately 10 years since the Oslo and Paris Conventions were negotiated. Being the first regional conventions of their kind, they developed in a piecemeal fashion. Subsequent regional agreements, such as the Helsinki Convention or the Regional Action Plans developed under UNEP's Regional Seas Programme, have tended to draw all aspects related to marine pollution, whether by dumping, from land-based sources or even pollution from the atmosphere, under one legal instrument. Together with the Bonn Agreement, which is an agreement between the North Sea States to combat pollution following an oil spill or incident involving hazardous chemicals, the North East Atlantic region is served by three separate organizations and legal regimes, and in addition there is the work of the EEC which clearly has a great impact in this field.

This organisational structure has both advantages and disadvantages. Certainly over the last 10 years the work has probably progressed more quickly because each Commission has had its own particular problem to deal with. A glance at all the rules, procedures and decisions which have been taken will confirm that much has been achieved. This division of the work into three compartments at international level also reflects the national organisations for dealing with marine pollution in many Contracting Parties.

However, as the preamble to the Oslo Convention notes, marine pollution has many sources and the Commissions do recognise that they are dealing with different aspects of the same problem. Therefore, there has been an increasing tendency for the Oslo and Paris Commissions to work more closely together on certain matters and they have established a common Secretariat. In recent years they have met together to discuss subjects of mutual interest, notably in the technical field matters related to the work of the Joint Monitoring Group. In 1984 the Commissions will meet in a joint session to take stock of their achievements over the past ten years and discuss their policy for the coming decade. It has been suggested that the time might be approaching when it would be appropriate to consider whether the interests currently served by the three conventions would best be served by a single convention.

It would not be honest to leave this review of the work of the Oslo and Paris Commissions without reference to one particular principle which is enshrined in both Conventions and on which it must be admitted that little progress has been made. This is the injunction that in drawing up measures, both individually and jointly, to combat or prevent pollution, the Contracting Parties shall 'harmonize their policies in this regard'. As is well known there is a fundamental difference in approach between the United Kingdom, following its EQO philosophy, and the UK's continental neighbours pursuing their limit values approach based on the 'best practical means'. This difference of philosophy was evident at the beginning of the 1970s and still pervades discussions at international level today.

In the case of the Paris Commission, it has during the past year attempted to carry out an assessment of the UES and EQO approaches to eliminating mercury pollution based on the different heads which were agreed by the Commission back in 1978: monitoring, reduction rates of total mercury loads, national measures to reduce mercury pollution from land-based sources other than the chloralkali industry, and emission standards set to respect quality objectives. After examination by the Commission's technical groups, the Commission came to the conclusion that it was not at present possible from the monitoring data available to compare the consequences of the two approaches for the marine environment. It was noted, however, that all Contracting Parties were taking technical measures to reduce mercury discharges. In the light of this knowledge, the Commission agreed that both approaches should continue to be permitted at least until a proper evaluation can be made at an appropriate future time.

Ch. 9]   The Role and Activities of the Oslo and Paris Commissions

It is the case that the different States which are Parties to the Oslo and Paris Conventions have different environmental problems and consequently have different solutions to those problems. In order that the administrations concerned can obtain a better understanding of these different individual circumstances, and a better appreciation of a country's environmental policy, the two Commissions have during the last two years set aside two days when each State in turn has made a presentation describing its national policies towards waste disposal with the object of putting the disposal at sea option in context. It is undoubtedly true that as a result of these presentations the members of the two Commissions now have a better understanding of each other's problems and philosophies. The experience gained will be valuable to the Commissions in formulating their policy for the next decade.

9.8 AUTHORS' NOTE

The authors would like to express their gratitude to the Chairman of the Oslo and Paris Commissions for permission to write this chapter. Any views expressed are those of the authors and do not necessarily reflect the policies of the Commissions. Any errors are entirely the authors' responsibility.

CHAPTER 10

European Community policy on environmental management

A. J. FAIRCLOUGH
Commission of the European Communities, Belgium

10.1 INTRODUCTION

To some it comes as a surprise to learn that the European Community has a policy on environmental management. It is not, after all immediately apparent what environmental protection has got to do with the operation of a common market.

This chapter will accordingly describe the origin, ten years ago, of the Community's Environmental Action Programme, indicate its *raisons d'etre* and achievements to date, comment more particularly on its effects on the field of water pollution control and indicate how the emphasis is changing with the adoption in 1983 of the latest Environmental Action Programme, for the period 1982–86.

10.2 THE ORIGINS OF COMMUNITY ENVIRONMENTAL POLICY

Article 2 of the Treaty of Rome speaks of the promotion throughout the Community of 'a harmonious development of economic activities and a continued and balanced expansion'. A qualitative as well as a quantitative approach is implicit in this and it was thus natural when, in 1972, at the time of the Stockholm conference the Community Heads of State and Government included the following statement in the declaration [1] made at the end of their meeting:

> Economic expansion is not an end in itself: its first aim should be to enable disparities in living conditions to be reduced. It must take place with the participation of all the social partners. It should result in an improvement in the quality of life as well as in standards of living. Particular attention will be given to intangible values and to protecting the environment so that progress may readily be put at the service of mankind

[Ch. 10] European Community Policy on Environmental Management 159

The institutions of the Community were asked to prepare an environmental action programme and thus Community environmental policy was born. The first Action Programme of the European Community in the field of the environment was approved in 1973 [2] and was updated in 1977 [3]. The term 'environment' has been broadly interpreted to include all natural resources which may be damaged or overexploited as a result of social or economic development; and also 'quality of life' considerations. The aim has been to ensure the sound management of natural resources and to bring qualitative aspects into the planning and organisation of economic and social development.

Another fundamental consideration underlying the Community environment policy was the recognition that if in response to the concern over environmental issues felt Member States were to adopt separate measures of a widely differing character, this would inevitably lead to distortions of competition, interfering with the proper functioning of the Common Market.

10.3 THE FIRST ENVIRONMENTAL ACTION PROGRAMME

It is important to note that the First Action Programme was negotiated on a 'line-by-line' basis through the Council of Ministers and adopted relatively rapidly and unanimously. The strong political commitment on the part of the Member States towards taking a wide range of measures to protect the environment was thus made clear. That commitment has not really wavered since, despite the recession and many other problems. One reason for this somewhat surprising fact must lie in the continuing (and even growing) public demand for adequate measures of environmental protection.

The First Action Programme set out the objectives and principles of Community environmental policy, and emphasised a number of basic approaches which remain important:

- prevention of pollution and nuisances is recognised as being better than cure;
- taking account of environmental needs at the earliest possible stage in decision-taking processes is stressed;
- the essential features of the 'polluter pays principle' are outlined;
- the principle of always determining the appropriate geographical level for any necessary actions was enunciated (thus, in particular, proposals for harmonisation at Community level should not be automatic but only made when action at Community level is needed in the light of the facts).

The First Programme went on to spell out in some detail the actions necessary for environmental protection. A large part of it was devoted to the reduction of pollution and nuisances. Priority pollutants were identified, overall approaches were defined, and details were given of measures needed relating to certain products, to certain industrial sectors, to certain areas of common interest (such as the sea and the river Rhine) and in relation to wastes.

10.4 THE RECORD TO DATE

What followed was a period of intense activity on the environmental front which has continued up to the present — particularly in relation to pollution control and prevention. In this action legislation has played a very significant part. Some 70 legislative measures have been adopted in the period of less than ten years that have passed since the First Programme was adopted. A good number of the measures adopted are concerned with water pollution. This initial emphasis on water reflected the priority given because of the characteristics and importance of water. It was chosen for priority treatment because it was considered to be the medium under the greatest threat by the disposal of industrial and domestic wastes.

Before looking in a little more detail at this water pollution field two general comments seem to me worth making. The first is that the fact that it has been possible in such a relatively short time to establish such a substantial body of legislation (supposed by numerous other actions in many fields of a non-legislative character) underlines strongly a point that I have already made viz. the high degree of political commitment on the part of the Governments of our Member States to effective action under the Programmes to protect the Community's environment. Of course there have been differences — sometimes acute — as to the 'how'. Sometimes there have been doubts as to the need for particular measures. Sometimes there have been disputes as to standards or approaches. But I cannot think of many areas of Community activity where so much has been achieved to build up a corpus of harmonised Community policy in such a relatively short time.

My second comment is that this simply could not have happened unless the proposals put forward had corresponded to Governments' perception of what was needed. When the Council of Ministers comes to negotiate on the basis of such proposals the reality is that the 10 Member States (each of them with their own particular national interests to protect and each of them with a commitment to the Community and to the concept of Community action when this is necessary to secure the proper functioning of the Common Market or to secure the harmonised application of environmental standards, norms, quality objectives etc.) are involved in striking a balance between national interests (often very disparate) on the one hand and the sense of Community commitment and their judgement of the need for Community action on the other. Proposals made to the Council can fail for all sorts of reasons — although few of our environmental proposals have in fact failed. But they cannot succeed unless the necessary political will exists in the Council.

In general therefore I think that the Community's environmental action has

Ch. 10]   **European Community Policy on Environmental Management**   161

to be judged a success at the legislative level. It has broadly speaking responded to public and Parliamentary demands and has commanded the support of Governments. The implementation of directives by Member States is, of course, another question — but to discuss that adequately would need a separate chapter if not a separate conference.

10.5 ACTION IN THE FIELD OF WATER POLLUTION CONTROL

I have already mentioned the degree of priority given, *de facto*, to tackling water pollution problems in the early stages of Community environmental action. The legal background to the reduction and control of water pollution is provided by directives (and a decision) which are administered within each Member State by individual national legislation. At the present time most of the obvious sectors of water pollution are covered by such Community legislation but the Commission has a forward looking policy and I will later suggest where future legislation might be directed.

The existing set of directives which I will not merely describe — that has been done many times before — can conveniently be grouped according to how they are implemented in actual practice. I shall be concerned to show how they can be applied as flexible instruments rather than as a rigid set of international laws; and to remind you that they can be brought up to date to deal with a changing situation or to take account of improvements in scientific and technical knowledge.

Let us start then with the 'Dangerous Substances' Directive [4] the famous ENV 131. This provides an example of how control can be applied to specific substances or groups of substances by aiming for the elimination of pollution caused by List I substances and for the reduction of pollution caused by List II substances. The Directive is a framework directive and must be implemented by subsidiary directives for List I substances whereby authorisation for discharge is based either upon fixed emission limits or upon quality objectives — in both cases as established by the Council.

It is perhaps unnecessary to remark that these two approaches represent different philosophies — that really was the reason why there have been such long-running arguments over the Directive and its implementation. Perhaps more interesting is to note the fact that, in the provisions made for 'new establishments' in the two subsidiary Directives adopted so far (mercury from chloralkali works [5] and cadmium [6]) there has been, in a sense, a certain coming together of the two approaches.

Be all that as it may, it can I think be acknowledged that there are attractions to both approaches. The use of limit values to fix emission standards is more attractive from the administrative viewpoint and contributes to the intention to avoid unequal conditions of competition. The use of quality objectives to fix emission standards intends to make best use of the receiving waters.

Apart from mercury and cadmium, List I refers to families or groups of substances rather than to individual substances. With the assistance of a group of national experts, the Commission has identified 129 substances [7] for priority attention on the basis of their toxicity, persistence and bioaccumulation. A procedure has been established, the first stage of which is the preparation of an ecotoxicological study for consideration by scientific experts. If appropriate, technological and economic studies are prepared for further consultation which leads to the definition of limit values for emission standards or alternatively for quality objectives which are then incorporated into a proposal for a directive. From a technical and scientific viewpoint, this procedure is thorough yet flexible enough to be really adapted to any potential threat to the environment. It has, however, proved to be a lengthy process to make separate directives for each substance from an administrative viewpoint and methods are being sought to speed up the procedure.

So much for ENV 131. The Second Group comprises a number of directives which lay down quality objectives for water used for particular purposes. These include the 'Surface Water' [8], 'Drinking Water' [9], 'Bathing Water' [10], 'Freshwater Fish' [11] and 'Shellfish' [12] Directives. These Directives translate quality objectives into lists of parameters and appropriate numerical values. There is a need to modify the 'Surface Water' Directive to bring it into alignment with the more recent 'Drinking Water' Directive, by allowing for changes which take place during treatment. Other changes to directives in this group are only likely to be made after appropriate consultation in the light of scientific and technical progress.

The one decision adopted to date establishes a common procedure for the exchange of information on the quality of surface water [13]. Consultations are currently taking place on changes to the Decision to improve its ability to meet its objectives. These include assessment of long-term trends and improvements; and laying the foundation for a comparable system of monitoring freshwater pollution. Agreement is also being sought from Member States for the information to be incorporated into the Global Environment Monitoring System as part of the United Nations Environment Programme.

The third category of action relates to certain specific industrial processes that are known to introduce pollutants into the environment. Here the initial concept was to apply directives to the specific industries concerned as a means of reducing such pollution; but there is in fact only one Directive of this type — on waste from the titanium dioxide industry [14]. This Directive requires Member States only to grant authorisation for the discharge of wastes on the general condition that there shall be no deleterious effects on the aquatic environment or other legitimate users of the waters affected. There are also requirements on Member States with regard to monitoring; and with regard to the drawing up of programmes for the progressive reduction and eventual elimination of pollution. These programmes are currently being brought into effect; and the

Commission, as also required by the Directive has recently made proposals [15] for the harmonisation of Member States programmes.

There seems little likelihood of the adoption of similar directives for other industries in the near future. In 1975, the Commission did submit a proposal [16] for a directive on the reduction of water pollution caused by paper pulp mills. But the Council failed to reach agreement on the proposal and the draft directive has simply remained on the Council's table up to the present.

The general situation with regard to the protection of water quality is that legislation has been drawn up for all sectors of the aquatic environment, and for all uses of water, where problems were known or anticipated to exist. Reduction of pollution is being achieved by the implementation of directives and the improvement of directives according to the latest technical and scientific knowledge. However, this policy is not one of complacency, nor does the Commission consider that we have yet reached the end of the road — far from it.

The Commission pursues an active research programme and maintains an awareness of developments so as to be able to anticipate and cope with possible future problems. Elements which might be included in the future in a forward-looking policy (and which could, if necessary, be supported by future Community legislation) might include some or all of: measures for the control of new chemicals; the tackling of the problems presented by diffuse sources of pollution (such as agricultural and urban run-off and the contamination of groundwaters by nitrates); an appreciation of the long-term implication of the general pollution of groundwaters; the possibility of establishing a minimum level of quality for all Community waters; the question whether (especially in regions of shortage like the Mediterranean) more emphasis should be placed on the development and wise management of water resources; the continuing problem of pollution of the sea by hydrocarbons and the growing problem of such pollution by dangerous chemicals; and the necessity for developing a global approach to avoid the transfer of pollution from one medium to another. A long enough agenda, I think, to keep us busy for some time!

10.6 THE COMMUNITY'S THIRD ENVIRONMENTAL ACTION PROGRAMME (1982–1986)

Having thus, to some extent, speculated as to what the future might hold in the field of water pollution and water resources, let me return to the more general aspects of the Community's policy for environmental management and say something about the Third Environmental Action Programme, the general approach of which was approved by the Council of Ministers in a Resolution adopted in February 1983 [17].

The First Action Programme (and also the Second, adopted in 1977) focused largely, as I have explained, on pollution control; and on lists of specific actions to be undertaken in order to, so to say correct the errors of the past. The Third

Programme by contrast is much broader in its approach; and is much more concerned to establish a general framework, a general approach, aimed at the positive management of resources so as to meet the environmental needs of the future. The key element in that approach is clear — for the future the 'name of the game' is prevention.

There has been a real shift of emphasis. The Third Programme states flatly, right at the beginning, that 'the resources' of the environment are the basis of — but also constitute the limits to — further economic and social development'. So the Community (and this is not just the Commission — it is the Council of Ministers as well) is saying that the environment is fundamental. Indeed the programme goes on to say that 'environmental policy is a structural policy which must be carried out without regard to the short-term fluctuations in cyclical conditions, in order to prevent natural resources from being despoiled and to ensure that future development potential is not sacrificed'. The only justification for such a view in a time of economic difficulties is that prevention is now seen to be cheaper than cure — certainly in the long-term; and often in the short-term as well.

Against this background, the commitment to a preventive appoach is seen as requiring the integration of environmental considerations into the planning and implementation of actions in virtually all other important fields — agriculture, industry, transport, energy, regional policy, tourism, overseas development, to name only those areas specifically mentioned in the Programme. Two examples, in particular, are stressed; illustrations of what is meant by this — the so-called Sixth Amendment [18], a Directive which establishes a pre-marketing system of notification and hazard assessment for all new chemical substances, and the Commission's proposed Directive [19] (which is still under discussion by the Council) on the environmental impact assessment, prior to authorisation of all major development projects likely to have a significant impact on the environment.

Moreover considerable emphasis is also placed in the Third Programme on the wise management of resources — land (especially ecologically sensitive zones), flora and fauna, water resources. Action to secure the better management of waste (reducing waste arisings, increasing re-use and recycling, and the safe disposal of unavoidable remaining wastes) is seen as necessary and, to this end, Community support for the development of clean or low-waste technologies is needed.

Another point of some importance in connection with the Third Programme and the way that we are tackling work under it, relates to the emphasis it places, as it must, on the socio-economic problems of the day. It says that environment policy must be managed in such a way as to try and 'contribute to the efforts made in other ways to find solutions' to those problems — the problems of unemployment and of economic recession.

There is quite a lot of evidence that certain environmental actions (just think of the construction of sewerage or of sewage works, or the construction of by-passes — all of which may be regarded as environmentally necessary) are considerable employment generators. And a lot of studies have been done which suggest that on balance, increasing the stringency of environmental requirements can in fact be a generator of employment.

Now of course I know that that does not produce the money to finance the necessary investments — and that is a point that is always made to us by industrial companies who say that Community directives are imposing costs upon them. On that too, however, there are some quite interesting figures which demonstrate that in some cases, at any rate, it is perfectly possible to tighten up on, in particular, pollution control requirements and at the same time for companies to generate increased profits; one cannot always hope for this but there are such cases. Moreover, there are companies, both in this country and elsewhere, who have adopted as part of company policy, an approach which involves taking a really very stringent view of environmental requirements — simply because they think that is good business in the long run.

10.7 THOUGHTS FOR THE FUTURE

In this area of encouraging people to believe that prevention is actually cheaper than cure in the environmental field, quite a lot of interesting things come together. For example, with a great deal of industrial restructuring going on, old industries are dying. Some of the new industries involve innovative technologies and many of those technologies are, by their very nature, less polluting and less environmentally troublesome than some of the old ones that they replace. Moreover, some of the innovative techniques can be and are being used in what I might call environmental management industries — pollution control equipment and so forth — and there is, in a sense, a potential coming together of improved environmental management with techological innovation, provided that people are ready to look far enough ahead and are capable financially of looking far enough ahead. So part of our business in the industrial field will be to try and persuade people that that does make good sense.

In this connection it is obvious that market forces have a very important part to play; and, in that context, the whole question of economic instruments is very important. The Commission is studying these and considering how the framework of economic judgement could be adjusted so that what is perceived as economically desirable by the investor, by the man who is taking the decisions, is actually the thing which will produce the environmental results that one seeks. In other words could we structure the economic framework better so that we avoid a situation of conflict between economic goals and environmental goals; and arrive instead at a situation in which what is economically sane and sensible actually leads to the environmental results we want?

Finally let me return from these rather general reflections to a concrete and urgent environmental problem that will, I believe, be high on the Community's agenda for some years to come — that of air pollution. The dying back of forests is beginning to happen in many countries, both in the Community and outside; emissions of, in particular, sulphur and nitrogen oxides (for example from combustion and other industrial installations and from motor vehicles) are thought to be implicated in this phenomenon, although the science is not yet clear; and it seems very likely that rapid action will need to be taken at Community level. The Commission has already proposed one Directive [20] (on emissions from industrial installations) and will be proposing others. The Commission is also, at the Council's request, urgently considering the case for lead-free petrol, which is a linked issue.

It is natural that this issue of air pollution has come under discussion at the highest political levels. At the Stuttgart summit meeting in June 1983, the Heads of State and Government, for the first time since the original summit of 1972, put a whole section in the conclusions of the meeting on the environment in very strong terms [21]. They referred to 'the urgent necessity of accelerating and reinforcing action ... aimed at combatting the pollution of the environment'; they referred in particular to 'the acute danger threatening the European forest areas' and called for 'immediate action' to avoid 'an irreversible situation'. This is the sort of language that one does not normally find in the conclusions from a summit meeting; so I think that the meeting, particularly with reference to air pollution (and also to lead in petrol) has put the environment right in the centre once again, in a way that it has not been (at that sort of level) since 1972 when the Community's environmental action all started.

I can perhaps conclude, on that note. The high importance of a sound and forward-looking environmental policy as a key element in the whole body of the Community's socio-economic policies is once again clearly seen and recognised. The priorities have changed — ten years ago it was water pollution; now it is air pollution; tomorrow perhaps waste management — to reduce the huge waste of resources that is involved in the situation as it exists today in the Community. But whatever the priorities the central message remains the same: it really does not pay to ignore or set aside the needs of the environment. They will catch up with us sooner or later; and when they do the bill will be much heavier than if they had been properly dealt with in the first place.

10.8 REFERENCES

[1] Declaration at the meeting of Community Heads of State and Government, Paris, October 1972.
[2] Declaration of the Council of the European Communities and of the Representatives of the Governments of the Member States, meeting within the Council, of 22 November 1973 on the implementation of a European

Community programme of action on the environment (Official Journal No. C 112, 20.12.73, p.1).

[3] Resolution of the Council of the European Communities and of the Representatives of the Governments of the Member States, meeting within the Council, of 17 May 1977 on the continuation and implementation of a European Community policy and action programme on the environment (Official Journal No. C 139, 13 6.77, p. 1).

[4] Council Directive of 4 May 1976 on pollution caused by certain dangerous substances discharged into the aquatic environment of the Community (Official Journal No. L 129, 18 5.76, p. 23).

[5] Council Directive of 22 March 1982 on limit values and quality objectives for mercury discharges by the chlor-alkali electrolysis industry (Official Journal No. L 81, 27.3.82, p. 29).

[6] Council Directive on limit values and quality objectives for cadmium discharges (adopted at Council meeting on 16 June 1983 – not yet published).

[7] Communication from the Commission to the Council on dangerous substances which might be included in List 1 of Council Directive (Official Journal No. C, 176, 14.7.82, p. 3).

[8] Council Directive of 16 June 1975 concerning the quality required of surface water intended for the abstraction of drinking water in Member States (Official Journal No. L 194, 25 7.75, p. 26).

[9] Council Directive of 15 July 1980 relating to the quality of water intended for human consumption (Official Journal No. L 229, 30.8.80, p. 11).

[10] Council Directive of 8 December 1975 concerning the quality of bathing water (Official Journal No. L 31, 5 2.76, p. 1).

[11] Council Directive of 18 July 1978 on the quality of fresh waters needing protection or improvement in order to support fish life (Official Journal No. L 222, 14 8.78, p. 1).

[12] Council Directive of 30 October 1979 on the quality required of shellfish waters (Official Journal No. L 281, 10.11.79, p. 47).

[13] Council Decision of 12 December 1977 establishing a common procedure for the exchange of information on the quality of surface water in the Community (Official Journal No. L. 334, 24 12.77, p. 29).

[14] Council Directive of 20 February 1978 on waste from the titanium dioxide industry (Offical Journal No. L 54, 25 2.78, p. 19).

[15] Commission Proposal for a Council Directive on procedure for harmonizing the programmes for the reduction and eventual elimination of pollution caused by waste from the titanium dioxide industry (Official Journal No. C 138, 26 5.83, p. 5).

[16] Commission Proposal for a Council Directive on the reduction of water pollution caused by wood-pulp mills in the Member States (Official Journal No. C 099, 2 5.75, p. 2).

[17] Resolution of the Council of the European Communities and of the Representatives of the Governments of the Member States, meeting within the Council, of 7 February 1983 on the continuation and implementation of a European Community policy and action programme on the environment (1982–86) (Official Journal No. C 46, 17 2.83, p. 1).
[18] Council Directive of 18 September 1979 amending for the sixth time Directive on the approximation of the laws, regulations and administrative provisions relating to the classification, packing and labelling of dangerous substances (Official Journal No. L 259, 15 10.79, p. 10)
[19] Commission Proposal for a Council Directive concerning the assessment of the environmental effects of certain public and private projects (Offical Journal No. C 169, 9 7.80, p. 14).
[20] Commission Proposal for a Council Directive on the combating of air pollution from industrial plants (Official Journal No. C 139, 27.5.83, p. 5).
[21] Conclusions of the presidency on the proceedings of the European Council, Stuttgart, 17–19 June 1983.

CHAPTER 11

The approach of the Federal German Republic – anticipating environmental effects

PROFESSOR J. SALZWEDEL
The Council of Experts on Environmental Matters,
Federal German Republic

Among the uses of surface water, considerable importance attaches to the extraction of water for industrial water supply and industry on the one hand and to discharge of municipal and industrial effluent on the other. However, these two main types of use are to some extent uneasy bedfellows inasmuch as the closer the relationship of effluent disposal to public water supply, the greater the problems to which effluent disposal gives rise. Central to management of water resources by the authorities is the task of regulating effluent disposal in such a way as not to place the public water supply at risk. Up to promulgation of the Fourth Amendment to the Water Resources Policy Act only the public water supply received special recognition in Federal law: according to Section 6 of the Water Resources Policy Act, authorisation and permission are to be refused if the envisaged use is prejudicial to public well-being, particularly it is is expected to pose a threat to the public water supply. On the other hand it was left to the discretion of management to ensure that the requirements for proper effluent disposal were duly met.

1976 was important for elaboration of the Water Act. In that year, partly from political conviction, partly through far-reaching adoption of provincial legislation or, at any rate, administrative practice agreed within the association of provincial water concerns, the Federal legislator conceived a Special Effluent Act which has, as it were, been grafted on to the established water management law.

(a) The new structure of effluent disposal legislation acquires the force of Federal law with its Fourth Amendment whose primary provisions are as follows:

- The introduction of the Anticipation Principle in water law, i.e. the acceptance of purification requirements that are not specific to a given waterbody, nor based upon utilisation of that particular waterbody.
- The associated levy of an Effluent Tax.
- The introduction of liability for effluent disposal incumbent upon public institutions, and effluent disposal planning.
- Ad hoc application of the Water Law to indirect dischargers, i.e. users of municipal or associative drainage systems.
- Partial overlapping of the Water Law by the Waste Law: only what in accordance with Section 1 Clause 3 Para 5 'Waste 6' is understood under the special meaning of the Effluent Law to be 'effluent insofar as introduced into waterbodies or sewage plants' will be excluded from the scope of application of the Waste Disposal Act, whereas otherwise the introduction of liquid wastes will be governed by both the Waste Law and the Water Law.

MINIMUM REQUIREMENTS FOR PRODUCTION AND PURIFICATION TECHNOLOGY IN ACCORDANCE WITH SECTION 7a OF THE WATER RESOURCES POLICY ACT AND EFFLUENT TAX

While all management requirements incumbent on the users of surface waters must be based specifically upon the surface water concerned, that is, must be governed by the utilisability and loading capacity of the receiving water concerned, the Effluent Law lays down discharge standards which must be observed without regard to the state of the surface water in question.

The States have indeed long insisted that all effluent discharges observe the requirements of accepted practice which were formulated first as standard requirements and later as standard values for sewage purification. The application of such discharge standards pre-supposes nevertheless that the necessity for them was justified in each case with specific reference to the waterbody concerned. Because of increasing surface water pollution this sometimes led to difficulties as it could well be argued that no higher grade utilisation could in any case be made of such an overloaded receiving water.

The Anticipation Principle is, in the first place, described in Section 1a Clause 1 of the Water Resources Policy Act as a management function of the water authority for the reason that any action prejudicial to the surface water must be avoided regardless of whether presently practised utilisations are distributed or critical water conditions reached. Section 1a Clause 2 of the Water

Resources Policy Act also establishes a general civic 'policing' duty: every person is obliged, in connection with measures that may affect the surface water, to exercise every care that the circumstances demand to avoid pollution of the water or any other adverse change of its properties. This applies regardless of the utilisation then being practised and the degree of loading of the water. This regulation becomes even more important when the water authorities, supported by the general 'policing' clause intervene in accordance with Section 3 of the Water Resources Policy Act against a threat to the waterbody below the actual utilisation threshold. It follows therefore that avoidable prejudicial actions can only be prohibited because this benefits the quality of the waterbody.

The rule for implementations of the Water Law is given in Article 7a of the Water Resources Policy Act: a permit for discharge of effluent may only be granted if the quantity and noxiousness of the effluent is kept as low as is possible when employing the process concerned in accordance with generally accepted engineering practice. From this are derived minimum requirements for quality of the discharged effluent and these must be included in any permit issued. The indefinite Federal legal concept of 'minimum requirements' laid down in Section 7a Clause 1 Sentence 3 of the Water Resources Policy Act is quantified by administrative regulations for 'interpreting standards' which must be decreed by the Federal government with agreement of the Bundesrat (Federal Council).

(b) The minimum requirements are, *ipso facto*, coupled with two legal consequences:

- Any permit for effluent discharge must be so limited by its content or qualified by such conditions or instructions that the pollutant loading of the effluent corresponding to the minimum requirements shall not be exceeded. For effluent discharges already existent in 1976, Section 7a Clause 2 of the Water Resources Policy Act grants the Provinces wide discretion to bring their permits gradually into line the minimum requirements.
- Whosoever observes the minimum requirements need, in accordance with Section 9 Clause 5 of the Effluent Amendment Law, pay only half of the Effluent Tax for which he is liable. However, should the permit be attended by stricter requirements for reasons specific to the water concerned, halving of the tax will be contingent upon observance of the said stricter requirements.
- Later elaboration of the Water Penalty Law in Section 324 of the Penal Code additionally shows that the minimum requirements are attended by a further and probably most far-reaching sanction which had not been actually envisaged when the laws were amended: whosoever transgresses the minimum requirements runs the risk of committing the statutory offence of unauthorised pollution of water.

In this connection it is hardly possible to trace the path of erroneous reasoning which led us to our present false position. Still, be that as it may, these so-called monitoring parameters which serve to quantify the minimum requirements offer an open sesame for intervention by the State Attorney for the Environment who is nowadays the most important protector of our water resources.

From Section 7a Clause 1 of the Water Resources Policy Act it is immediately clear that the minimum requirements are not designed to warrant notices of objection. The rules of generally accepted engineering practice cover only a relatively modest technical level and can thus in principle be observed by any user of water, according to his nature, without any great difficulty. For some rules to be considered as belonging to accepted practice, they must be known and accepted as right and proper in the relevant engineering circles. They must then have been tested in practice, proved satisfactory and been adopted by the majority of specialists working in the technical field concerned. It therefore depends only on the convictions of the sewage engineers who are chiefly engaged in the treatment of municipal or industrial sewage or, more precisely, on the majority opinions they have formed. The reasons to which a delayed acceptance of a long current technique can be attributed are irrelevant in law. In any case the subsequent adaptation of an old-established utilisation to the minimum requirements can in some cases give rise to technical or spatial problems. However, the cancellation of such an old-established permit would not be justified on grounds alone of the impossibility of subsequent adaptation but only for reasons specific to the water concerned. It must also be emphasised that the minimum requirements specified in permits do not by their nature offer any protection to neighbours.

Since the regulations bring into force the Anticipation Principle and hence are not directed at the attainment of any defined water quality, subjective rights of the neighbour under Water Law cannot be affected and the sole criterion is the common good. This also serves to allay the exaggerated disquiet which is still felt about agreements between government and branches of industry or statutory agreements between the authorities and an individual concern regarding the extent of purification instructions. As a rule this concerns agreements on emission reduction, not the pollution load, which is expected of other water users. Should third parties complain about the agreements or their application in individual cases they will be thwarted by the right of action. It would incidentally be an error to believe that the introduction of Federal legal discharge standards can be coupled with the adoption of anticipatory policy concepts embodied a short time previously in Section 5 Para 2 of the Federal Emission Protection Law. The more perceptive observer might consider that the Anticipation Principle was first adopted in the Federal Emission Protection Law with strict standards in accordance with the state of the art and then in the Water Resources Policy

Act with moderate standards according to generally accepted engineering practice. But such an observer must realise that air pollution is worse, since air breathed cannot be purified even in conurbations or heavily industrialised areas, whereas water pollution cannot reach the consumer directly because, fortunately, purification plants bar the way. The introduction of Section 7a of the Water Resources Policy Act was intended to standardise instructions for purification in all States of the Federal Republic, with biological purification being generally and primarily employed, and also to afford the legal safeguard that experience has gradually shown to be necessary for purification instructions not formulated specifically for the water concerned. In retrospect a system of minimum requirements in accordance with Section 7a of the Water Resources Policy Act is indeed also based — and rightly so — on the Anticipation Principle: in accordance with Section 1a of that Act it is intended that avoidable pollution loads be ruled out and surface waters remain as, it were, as clean as possible. To suggest that a yawning gulf exists between measures for air pollution control and water conservancy is incidentally quite erroneous. The emission limits in accordance with Section 5 Para 2 of the Federal Emission Protection Law is based on a conventional state of the art and not — as made clear by Section 3 Clause 6 of that law — on the absolute peak value. Implementation of the above in industry is to be governed by Part 3 of the TA-Air which is now next in line for review. Feldhaus too has always pointed out that the measure of progressiveness assigned to the art depends upon extremely complex processes of evaluation. Perhaps the difference between emission limitation for air pollution control and that for water conservancy could best be described as follows: the significance of state of the art for waste gas scrubbing has been laid down without accepting as a necessary precondition that a majority of specialists in this technical field should also have accepted this process; the minimum requirements according to Section 7a of the Water Resources Policy Act are, on the other hand, based on the rules of accepted engineering practice that are generally recognised and, in other words, accepted by the majority of specialists working in that field. Whether this really means much in practice is arguable and can hardly be proved one way or another. Examples ought to be given in which technologies are based on administrative regulations according to Section 7a of the Water Resources Policy Act that are more exacting than much of what is written on emission limitation in conection with air pollution control.

(c) The relation of Section 7a to Section 1a of the Water Resources Policy Act is debatable: can the possibility of technically avoidable water pollution loads be precluded over and above observance of the minimum requirements? Essentially Section 7a Clause 2 of this Act is a *lex specialis* and any further requirements must be justified specifically for the water concerned. However, if old-established installations covered by Section 7a Clause 2 of the Water Resources Policy Act are technically more efficient, Section 1a Clauses 1 and 2 of that Act will be determinant because the 1976 amendment was certainly not designed to do away with any existing capacity. The same must then surely apply in the future in cases where sewage works could be run with greater efficiency for the reason perhaps that they are not being fully utilised; in such cases the available technology must also be employed. Consequently, these requirements do not presuppose that the water authority offer some water-specific justification in addition to the proof of avoidability. However, since Section 1a of the Water Resources Policy Act cannot be intended indirectly to intensify the provisions of Section 7a Clause 1 of the Act, the question as to whether the mere presence of avoidance capacity should involve the need for its complete utilisation must be weighed up with particular care and very strictly on the principle of proportionality. Finally, it is necessary to examine the question — indissolubly linked to the Anticipation Principle — as to whether and to what extent the minimum requirements should be measured on the proportionality principle for the sector as a whole (for example, for the pulp industry) or for individual cases (for example because of the significance of a works for an undeveloped area). On the one hand it is clear that effluent purification will not be carried out just for its own sake but, as shown by the relation between Section 7a and Section 1a Clause 1 and 2 of the Water Resources Policy Act, in order to prevent the impairment of surface waters, pollution of the water or any other adverse change in its properties. On the other hand, the Anticipation Principle is directly designed to divorce purification requirements from considerations of emission, that is to separate discussion on water conditions from those on loading capacities. The objection that comparability does not exist is only relevant when the economic effort for reduction of pollutant loading appears quite definitely pointless because it can do nothing for the public good; it does not necessarily depend on the measurability of water quality improvement. The reduction of pollutant loading with deleterious substances, especially those on the black list, is always a justified requirement. Otherwise, with regard to the individual pollutant parameters it is a management policy task to provide storage and buffer capacity particularly in order to regain latitude for new permits. Meanwhile, there is a further point that merits attention in the Water Law: decades of implementation have clearly shown that technical requirements for waste production and sewage purification can only be enforced through identically formulated requirements for all dischargers without consideration of effects on the environment. In this connection Section 7a of the Water Resources Policy Act only confirms State practice. The levy of Effluent Tax attendant on implementation of the water legislation now ensures that the benefits of competition will be creamed off by taxation in cases where the technical requirements concerned cannot perhaps be justified on grounds specific to pollutant or area.

(d) The establishment of minimum requirements is based firstly on generally accepted rules for the production process and secondly on those for final clarification of the collected effluent. In between, there will often be special pretreatment for certain effluent flows. The minimum requirements make the generally accepted rules on which they are based neither directly nor indirectly binding. It is up to the waterworks itself whether to observe the minimum requirements by particularly advanced production processes, by extensive purification of part-flows or by particularly efficient final clarification of the collected effluent; in certain circumstances it is then accordingly permissible to fall below the required minimum state in the other process stages. On water management grounds, whether because of particularly sensitive utilisations or because of a particular preload, water legislation may even require full utilisation of advanced processes in all process stages, particularly in production.

The fact that co-operation between State and industry over administrative specifications to section 7a of the Water Resources Policy Act has led to more than application of a mere end-of-the-pipe technology can be regarded as a marked step forward. Both through discussion on production processes amicable to the environment and also on the separation and pretreatment of partial waters which the central sewage works cannot handle satisfactorily, it has become possible substantially to improve the overall cleaning result. All of this would have been inconceivable if one had remained content with implementation of generally recognised rules of effluent technology; nor even with the minimum requirements for the last attainable cleaning result would this have been done were it not primarily for the positive contribution made by the legal sequelae of halving the tax. Now the specifications contained in the Effluent Tax Law will also have to be reviewed in the light of improved knowledge: in fact the tax basis is the discharge of effluent into a body of water, but the determinant for tax calculation is the noxiousness of the effluent per Section 3 of the Effluent Tax Law in the form considered for partial waters and not as conceived of after intra-works dilution. It is especially important for partial effluent discharges to be measured within the works. Special monitoring parameters must then, in any case, be established for partial effluent discharges if these are channelled separately and direct to the body of water and therefore do not pass via the central sewage works.

The administrative specification on mixed water which is especially important for the chemical industry, should therefore not be regarded as typical. In this case, experience also shows that once extremely diverse effluent streams have been mixed, only limited purification results can be achieved and then only at great expense. If partial streams have to be subjected to a special regime in order to guarantee better results by taking preventive action, this is undoubtedly a fundamentally sound practice.

Just where the threshold for the partial flow concepts lies can be discussed only in each individual case.

The Federal Minister has set up 60 working groups for drawing up drafts of administrative specifications on minimum requirements per Section 7a Clause 1 Para 3 of the Water Resources Policy Act. Most municipal effluents are covered by the first general administrative specification on minimum requirements for the discharge of foul water from the municipalities into the receiving water of 24th January 1979 (1. Schmutzwasser VwV). In it, the minimum requirements for sewage works are graded according to date of building and size category. The so called monitoring parameters are those laid down for settleable substances, CAD and BOD_5. In the meantime, the function of the monitoring parameter has been re-amended. Originally, a monitoring parameter was regarded as not met if the arithmetical mean of the results of the last 5 determinations of this parameter as part of the State Water Inspection exceeds this value, determinations dating back more than 3 years are disregarded. In the meantime, there is a swing towards the following course of action: 'A parameter determined in No. 2.1 is met. It also counts as satisfactory if the arithmetical mean of the result of the last 5 investigations performed as part of the State Water Inspection does not exceed this value'. Considering the matter dogmatically from a purely legal point of view, it is tempting to insist that the minimum requirements be laid down in a statutory ordinance chiefly because not only instructions but also taxes are concerned. But this would only be disadvantageous from the dual standpoint of implementation and environmental assessment. As in all technical processes, and particularly where biological processes are concerned, deviations from technical rules must be admitted in atypical individual cases and anyone who prescribes a formal dispensation in this connection is merely throwing a spanner in the works. Otherwise the only moot point is whether the existing official authorisation still goes far enough. In order to use this or that generally accepted rule of engineering practice as a basis or to justify this or that minimum requirement it was desired, in the final event, to resort to the theory of essentiality in order to make it appear that when establishing the main technical points Parliament also accorded due weight to environmental considerations. This excessive demand would merely be embarrassing to everybody. Neither would combining the highest legal status with the minimum expert knowledge be the statutory *non plus ultra.* The reservation made in the law suffices to ensure that any provision in the governing administrative regulations can be amended at any time by the legislator or issuer of the ordinance. Anyone desirous of obtaining a legal opinion covering the given technical conditions must remember how different was the starting position in the administrative regulations which laid down the minimum requirements in accordance with Section 7a of the Water Resources Policy Act:

The Approach of the Federal German Republic

(i) From the purely legal standpoint the most favourable case would be to have a generally accepted rule of production technology and generally accepted rules for sewage engineering otherwise introduced by the States in accordance with Section 18b Clause 1 Sentence 2 of the Water Resources Policy Act established in a set of technical regulations. In the long term even reassessment of the minimum requirements cannot be elaborated on the basis of such an assured method for the formulation of objectives.

(ii) In the communal sector resort could already be made to approved and passed technical regulations. Admittedly, it transpired that the working party was not disposed to accept uncritically the processes described therein and use them as a basis for formulating the minimum requirements. The need therefore arises, after issue of the administrative regulations for foul water, to re-adapt the Sewage Engineering Association's (ATV) effluent regulations to the minimum requirements. But this in no way alters the fact that the preliminary work of the ATV has proved a valuable aid not only for passing of the administrative regulations on foul water but also for that of other administrative regulations as well.

In the case of the DVGW (German Gas & Water Association) and the ATV, regulations are produced by a similar procedure: a committee of the appropriate professional association prepares a draft regulation which is then published (in the case of the ATV as a yellow paper) to which objections can be raised within a certain time-limit (for example in the case of the DVGW usually three months) prior to the final version being ratified. This procedure is formally laid down, albeit with very variable frequency of adjustment. These sets of regulations are kept updated in step with technical development, this being assured in the case of DIN standards for instance by checking every five years.

The procedures described (publication of drafts and the opportunity to raise objections) enables the *sets of regulations* to be agreed by the specialists concerned but this does not necessarily imply that the measures of processes there described will be classified as 'accepted rules of engineering practice'. In the legal sense this depends only on the above mentioned criteria. When clarifying whether a formally approved rule belongs to 'generally accepted rules of engineering practice', 'the state of the art' or 'the state of scientific and technical advance' procedural regulations or instructions to users are important starting points. They indicate (unfortunately not always entirely clearly) what level of technical requirements should, in the opinion of the compilers, be represented and confirmed by their specialist colleagues in the sector concerned. Unambiguous in this connection is the hitherto valid ATV instruction to users stating: 'The set of ATV effluent regulations is ... rules of engineering practice'. However, given the wide range of possible technical processes, it can in individual cases depend upon the technical sectors concerned, and also on the powers of persuasion of the committee members and their ability to put their views across, as to whether a guideline shall be classified as belonging to 'accepted rules of engineering practice' or to 'the state of the art'.

The use of technical regulations for implementing the indefinite legal concepts of 'generally accepted rules of engineering practice', 'state of the art' or 'state of scientific and technical advance' do not presuppose in individual cases that they have been officially recognised whether in law or ordinance (as in the second and fourth implementation ordinances to the Water Law) whether in an administrative regulation or promulgated in some other manner (as for instance through the standards agreement of 5th June 1975 between the Federal Republic of Germany and the German Standards Institute (DIN)).

(iii) If for a given project a 'generally accepted rule of engineering practice' cannot be established by reference to an administrative regulation or a technical regulation, such rules must be devised from practical observation. 'Anticipated sets of regulations' can only be established when an essentially uniform picture has been built up from individual practical observations. That is the case when the official requirements which are regularly imposed when granting new permits to effluent dischargers and which cover processes accepted by the parties concerned as being feasible in practice. Starting points can also be offered by a draft set of technical regulations. Under no circumstances, however, does the inclusion of a technical rule in a set of regulations or an administrative instruction give grounds for its classification as 'generally accepted'. It may well be that certain processes or measures correspond to a generally accepted state of the art but that no rule yet been formulated whether because the responsible committees have not yet had the opportunity to formulate any requirement or because agreement had not been reached or because an earlier publication requires to be technically revised.

The ATV has so far prepared technical rules only for the purification of domestic sewage. A set of rules for industrial sewage treatment is in the course of preparation. For the time being therefore, the industrial sector must rely entirely on practical observation.

(iv) More difficult is the determination of generally accepted rules when comparison of practical examples fail to give a uniform picture. The situation may arise for instance where new permits are required for very diverse technical processes or where very different purification performances may be achieved with one and the same process. In such cases there is sometimes a danger that 'generally accepted' may be taken to apply to the rules of engineering practice representing the

lowest technical level of the process and measures compared. But conversely, one may very easily be induced to describe as 'generally accepted' that which has been well proven in individual works that are particularly go-ahead. This would mean upgrading the yardstick applied in Section 7a of the Water Resources Policy Act to 'state of the art'.
(v) The possibility cannot be precluded of a committee undertaking publication of administrative instructions on minimum requirements in an industrial sector, specifying measures for sewage purification that are hitherto unknown or at any rate have not yet been put into practice. In this case the observation of management practice will not only give no uniform picture, it will give no picture at all. A guide to possible sewage purification processes can be given by the experience of advanced concerns abroad, and comparative evaluation of technical intra-works measures in allied sectors or research results from university institutes. Last, but not least, account should be taken of tenders from companies who develop and market sewage treatment equipment.

As in the cases previously considered, the administrative instructions must be careful not to require too much or too little. Should they content themselves with the first modest beginnings of a suitable sewage treatment process their requirements level will lag behind that of the other working parties for other sectors. If the minimum requirements are set too high because one aims from the start at what is technically possible and because, in default of an adequate range of observations, the modifying characteristic of general acceptance fails, the industrial sector in question will be overloaded. The only way out probably is to examine all technically feasible processes before deciding whether to upgrade or downgrade.

It has been found that the text of Section 7a Clause 1 of the Water Resources Policy Act promises more than it gives for execution. With regard to generally accepted rules of engineering practice, the law makes compulsory a requirements level which has developed where such generally accepted rules of engineering practice already exist. For all practical purposes this holds good only in the case of processes for the treatment of domestic sewage. When seeking the right scale of minimum requirements for the purification of industrial effluent it must be a question of finding the right requirements level. The inhomogeneity of the picture offered by the range of industrial observations must neither induce too abundant a pollutant load on the water nor precipitate recourse to 'state of the art' yardsticks, which can easily be done when reviewing what is technically feasible.

The really important thing is that purification instructions should be devised not only from the standpoint of Section 7a of the Water Resources Policy Act but also with due regard to the receiving water concerned and if necessary to other more far-reaching requirements specifically applicable to that receiving water. The importance of all 60 working parties engaged in preparing the administrative instructions working to a comparable requirement level does not derive primarily from the need for all dischargers of a given industrial sector to receive identical purification instructions. From the standpoint of water conservancy this would even be highly unsatisfactory. The first consideration is that all water users be given essentially the same opportunities for halving their Effluent Tax in accordance with Section 9 Clause 5 of the Effluent Amendment Law. It would be awkward if, as a result of the working parties using different requirement levels, one branch had this 50% reduction handed to it on a plate while others found it tantalisingly difficult to obtain.

CHAPTER 12

The UK approach – environmental quality standards

J. A. L. GUNN
Department of the Environment

12.1 INTRODUCTION

The purpose of this chapter is to state and explain the principles on which water-pollution-control policies are now founded in the United Kingdom. It also explores some of the interactions between them.

The protection of the aquatic environment in this country is based on four principles:

(i) A single authority both exercises control over all discharges and abstractions in each river catchment and estuary, and also provides the water services throughout the area of that river basin. This is the *principle of unified management*, which makes the management of the aquatic environment an intrinsic part of the management of the whole water cycle.
(ii) Discharges are required to meet conditions which are set by reference to the environmental quality objective propounded for the receiving water; this is the *principle of using environmental quality objectives* (EQOs), and it implies not only a qualitative statement of the water's functions but also the specification of the chemical characteristics which relate to that function, as well as relying on a specification of the chemical properties of discharges themselves.
(iii) The polluter should pay the cost of whatever measures are necessary to avert, reduce or remedy any pollution he causes, and the requirements he must meet are set by public authorities. This is the *polluter pays principle* (PPP).
(iv) In environmental matters, as in others the British Government participates constructively in the work of the European Community (EC) and complies with its requirements. I shall refer to this as the *EC commitment*.

12.2 THE PRINCIPLE OF UNIFIED MANAGEMENT

The first principle is that a single authority exercises control over all discharges and abstractions, in order that it can provide the water services, and manage the whole water cycle, in each river basin. The justification for this is that a river fulfils a variety of functions — as a drainage channel, as a source of drinking water, as a conveyor of waste; perhaps for fisheries, irrigation, and recreation; it functions too as a natural habitat. The interests of those upstream in the uses of a river are in conflict with the interests of those downstream, and the interests of different users of the same stretch are also liable to conflict. These conflicts are not likely to be resolved fairly and economically unless there is a single controlling authority; and the resolution of these conflicts is likely to be fairer and more economic if the controlling authority is also responsible for securing all the principal uses of the river's waters.

The basic and elementary conflict is between the use of a river for waste disposal and its use for any other purpose. So the first step towards unified management is to establish an authority responsible for controlling discharges throughout the catchment. In England and Wales, the completion of this stage is marked by the Rivers (Pollution Prevention) Act 1951, which, building on legislation of 1948 and earlier, provided as the standard administrative set-up bodies each responsible for one or more rivers, each with the duty of maintaining and restoring its rivers' wholesomeness, and each equipped with the essential power of attaching conditions to the consents which dischargers were required to obtain. The use of that power enabled these bodies substantially to advance the quality of rivers in England and Wales through the 1950s, 1960s and in to the 1970s. The pattern of the 1951 Act can be seen in many developed countries, and is still the pattern in Scotland.

However, in England and Wales the Water Act 1973 took an important step further, by unifying the bodies responsible for protecting the riverine environment with the public bodies responsible for providing water supply and sewerage; and these new unified authorities were also given responsibilities for natural conservation and for water recreation. This integration of all the water services into a single, purpose-built authority is unique in the world. It cuts across the administrative boundaries and jurisdictions which elsewhere remain paramount, and which constrain or prevent the unified management of water resources.

Without a unified pollution control authority, nations or local authorities who share a river can recognise one another's interests, and cooperate in separately regulating its use. But that is a poor substitute, both in economic terms and in environmental terms, for unified management in its fullest sense. The set-up in England and Wales has the further advantage that responsibility for policy and action in each catchment — including the setting and pursuit of EQOs for each body of water — can be effectively delegated to a regional level and, indeed, to a sub-regional level.

The fundamental administrative reforms of the 1951 and 1973 Acts have been modestly taken forward in the Water Act 1983, which has strengthened the water authorities' control over local authorities as their agents for sewerage purposes. Another current — but also modest — administrative advance is the implementation of Part II of the Control of Pollution Act 1974 so as to bring into the system of control various waters, such as previously undesignated estuaries. A third current development, under the 1983 Act, is the ending of the local government predominance in the membership of a water authority. Some members will still be desired from local government, though now appointed by the Secretary of State. But local accountability will be sought in a new way. Water authorities will be required to establish arrangements for consulting the interests of their customers and users. This is a counterpoise to the possible consequences of the 1973 Act's separation of the management of the water cycle and services from all other forms of public administration. The authorities' aim must be so to draw up and work the new consultative arrangements as to provide and prove a reasonable degree of public satisfaction with their policies, priorities and action.

The principle of unified management has thus been under development for a generation. It should now be generally accepted and stable. It has developed with, and has contributed to, with development of EQOs, as will emerge more fully.

12.3 THE USE OF ENVIRONMENTAL QUALITY OBJECTIVES (EQOs)

The second principle of United Kingdom policy for the protection of the aquatic environment is that the standards to which individual discharges are required to conform should be set by reference to an objective for the quality of the waters affected. The use of EQOs depends on setting and monitoring measurable standards for individual emissions, and on setting and monitoring measurable standards for the quality of the receiving waters. The latter are intended to be primary, and to govern the setting of standards for individual discharges. Quality standards for the receiving waters are intended to relate to their function: supporting a certain level of fish life, or plant life, providing water which, after treatment, will be fit to drink, providing a means of waste disposal, providing for abstractions, for agricultural use and so on.

The justification for this approach is that there are great differences in the ability of different watercourses, estuaries and seas to absorb, disperse or neutralise a given pollutant input without prejudice to such other functions as that water may have. If emission standards for effluents are set without reference to the capacity of the receiving waters to handle them, such absurd results could follow as permitting a chloralkali works on a stream that dries up in summer, or ignoring the presence of naturally arising cadmium in regulating cadmium discharges. If a plant which gives rise to large quantities of acidic waste is sited on a substantial estuary to secure good dilution and disposal of the effluents, and its owners accept whatever other penalties that location may impose, there is no compelling reason to limit its discharges to levels set by reference to non-environmental conditions. The aim is to protect and improve the environment, as found, and not to eliminate discharges and dischargers.

The principle of using EQOs has emerged and been formulated only quite recently. Early British legislation and practice appear to have concentrated on fixing tolerable limits to the pollution characteristics of discharges, and providing an administrative and technical system which could apply them in a practical way. The inevitable emphasis at first was on the emission standard for the individual discharge, and on setting these at values attainable by the technology of the day. That is an essential point, though it may now seem rudimentary.

Two factors take practice forward from this elementary level. First, those who set emission standards do so in order to regulate the quality of the environment and will inevitably want to give precision to that concept, differentiating and defining alternative levels of achievement. Second, they will be interested in the possibility that the quality of the environment can be progressively improved or can deteriorate to a lower standard. They will want to avert detioration and achieve improvements. Once the aquatic environment in a river basin is managed as a whole, the setting of EQOs is not just a theoretical exercise, but a preliminary to practical decisions and actions over a wide area and for a long time ahead. No doubt isolated glimpses and remarkable applications of these ideas can be found in the first half of the twentieth century, but they have only become practicable possibilities in this country in the last generation, and have only become a standard procedure for planning and action in the last decade.

It seems to have been no accident that the progress of EQOs, from a general notion of 'the wholesomeness of rivers' (in the words of the 1951 Act) to a clearly conceived principle and then to a standard procedure for regulating river quality, advanced in step with practical experience and progress in the regulation and enhancement of the aquatic environment, and also with the development of unified management from an elementary principle to a more comprehensively conceived and now fully operating institution, established by successive Acts of Parliament. The creation of regional water authorities, made possible and worthwhile the work which culminated in 1978, in the National Water Council's report 'River Water Quality: the Next Stage', which gave coherent and comprehensive expression to the theory and practicalities of EQOs, including the idea of progressing from short-term to long-term EQOs. And at the practical level river quality has been greatly improved by the fact that the body which sets the EQOs also operates the sewage treatment works and undertakes the investment programmes which principally determines whether those EQOs are met.

However, for all its merits, the EQO approach is not entirely painless. The criticism could arise in future that it carries a costly penalty of water quality standard setting for different contaminants and different water uses. The principle dicussed in the next section could help resolve that difficulty.

12.4 THE POLLUTER PAYS PRINCIPLE (PPP)

The principle that the polluter should pay the cost of averting or remedying his pollution (as defined by the competent public authority) is justified not only by considerations of equity but also in economic terms: anyone who has to bear the cost of eliminating or reducing pollution resulting from his activities has an incentive to reduce the cost of complying with what is required. PPP has recently been examined in some depth by the House of Lords' Select Committee on the European Communities, and it is therefore unnecessary to explore it in detail in this chapter. Their report and evidence bring out that the principle has been adopted in many developed countries; and — an unsurprising consequence — has been formulated, interpreted and applied differently from country to country. The definitions agreed, and the exceptions which have gained international acceptance, tend to conceal substantial variations in national practices.

PPP's main limitation, as a general principle of public policy, is that a government may want to make faster progress with a pollution control programme than the polluters individually can afford. Where that is the case other parties or the public at large must foot the bill. Another limitation is that it is effective only where polluters can readily be identified. The House of Lords Committee point out that in the UK we have also fallen short of the fullest possible implementation of the principle in respect of direct discharges.

In England and Wales a difficulty has arisen in the application of PPP which does not arise in Scotland or on the continent: by unifying the responsibility for pollution control with the responsibility of providing sewerage services and sewage treatment, we have taken the risk that the water authority will seek to save sewage treatment costs by letting environmental quality go, or achieve it by imposing unduly stringent conditions on other dischargers. There are three safeguards: first the integrity of public servants; second the fact that the water authorities' own discharges are consented by the Secretary of State; and third the watchful eye of the public, now to be focused, we intend, through the Consultative Committee arrangements to which I have referred.

PPP is an important general principle of UK policy for the control of pollution in the aquatic environment. Its applicability to other environments, and the fact that it has international acceptance distinguish and separate it from the two principles already discussed in this chapter: it is compatible with both the unified management principle and the EQO principle; but it is equally compatible with other administrative arrangements for pollution control and with other principles of setting emission standards. In as much as some exceptions to it are recognised internationally, and it is not regarded as an absolute principle, it is less powerful than it sounds. But the United Kingdom has been more rigorous in applying it than most countries.

12.5 THE EUROPEAN COMMITMENT

The fourth principle of British policies for the aquatic environment is our commitment, as a country, to the European Community (EC). In environmental matters, as in others, the United Kingdom is a constructive member of the community and we comply with its requirements. That affects our priorities and our actions in relation to the quality of aquatic environment.

The Treaty of Rome itself makes no specific mention of the environment and does not provide for the EC to play any role in environmental regulation. However, Article 235 provides for the Council, acting unanimously on a proposal from the Commission, to take powers not otherwise provided where they are necessary to fulfil an objective of the Community (as described in Article 2 of the Treaty). Under this provision three EC Environmental Action Programmes have been formulated and agreed by the Council covering the periods 1973–77, 1977–81 and 1982–86. The principal means of implementation is by the making of Directives, which can only be proposed by the Commission. Directives are addressed to Member States and are adopted by the unanimous decision of the Council; they are binding as to results to be achieved, in whatever way accords with the Member States' law and conditions.

So far, some 16 Directives have been made on matters relating to the management of the water cycle; there are 7 current proposals for directives, and we know of another 8 subjects on which the Commission is considering making proposals. These directives fall into a variety of different categories: they are concerned with:

Products which can pollute the aquatic environment (for example detergents)
The quality of supplies of *drinking water*
Restrictions on the application of *sewage sludge* in agriculture
Dangerous substances generally (Directive 76/464) and specifically (mercury, cadmium etc.)
The *quality of natural waters* allotted to particular purposes — for supporting fish life; for bathing; as sources of drinking water.
Emissions by a particular industry (for example producing titanium dioxide; woodpulp)

and there are related directives and decisions on monitoring, the exchange of information and so on.

Attention has tended to concentrate on the Directives concerning pollution caused by dangerous substances discharged into the aquatic environment; starting with the important and far reaching parent directive 76/464/EEC which sets a framework for Member States' action to eliminate pollution by substances which are singled out as particularly dangerous by reason of their toxicity, persistence and tendency to accumulate in bio systems (list I — the 'black list'); and to reduce pollution by other dangerous substances (list II — the 'grey list').

The directive requires discharges containing list II substances to meet an emission standard based on EQOs, which may be laid down at community level, (such as for the support of fish life and for sources of drinking water), but otherwise are to be laid down by Member States. For list I substances there are two regimes; a 'preferred regime' under which Member States cannot permit discharges with concentrations exceeding limit values which are laid down in 'daughter' directive; and an alternative regime under which Member States must control list I substances through EQOs laid down at Community level in daughter directives.

It was provided in 76/464 that until daughter directives governed them individually, dangerous substances would be regulated on list II terms. It was assumed that progress would be made with 'the preparation of quality objectives determining the various requirements an environment may meet bearing in mind its allotted purpose' (to quote from the first Environment Action Programme), and such progress would have enabled the list II regime to be applied extensively, but this has not happened. The intention was also that daughter directives, applying the preferred or alternative list I regime to individual substances, would issue in a steady succession, but there have been difficulties here as well.

Directives have to be agreed by all member countries and negotiations on the first two daughter directives (on mercury from the chloralkali industry, and on cadmium) were difficult and slow, because not all Member States could readily accept the practical application of the alternative approach as defined in the parent directive. To some delegations it seemed unfair, and also a distortion of the terms of fair competition, that the UK could escape from the application of limit values; and so could permit even new industrial plant to employ less than the best pollution control technology in circumstances where the immediate environment could safely absorb and dispose of controlled amounts of dangerous substances. To the UK, which alone has adopted the alternative approach, it was equally strange that in other countries the pollution control authorities could apply fixed emission standards without paying any heed to the different absorptive capacities of a regularly flushing estuary and a small seasonal stream. However, in June 1983 the second daughter directive was agreed; we in the UK hope that the formulae it contains (which show sufficient regard to our position, particularly on new plant under the alternative regime) will provide a model for dealing with the same problem as it arises on other particularly dangerous substances. If so, there is no reason why the Community should not make rapid progress with daughter directives taking dangerous substances individually or in groups. It will remain essential that the numerical values proposed for individual substances are scientifically based and attainable in practice.

We would also hope that with this major problem resolved, progress can now be made with Environmental Quality Objectives at Community level. The work which the Department has commissioned, and which the Water Research Centre has done, on environmental quality standards for chromium, nickel, zinc, copper, lead, arsenic and other substances is intended to provide a constructive input.

The differences of approach between the United Kingdom and the other EC countries appear to stem from three factors. First a few of their major rivers cross international boundaries; as a result unified management, as we know it, is impossible. Second, several continental countries are at that stage in tackling river pollution, where all improvements are desirable and their cumulative effect is not at the centre of attention. Thirdly the EC countries are concerned to ensure that pollution controls are not applied in a way which brings inequalities into the terms of economic competition. To us that seems unreal: we would regard the ability to use a copious environmental capacity, where it exists, as a natural locational advantage, doubtless offset by other locational advantages at other places. However, these unavoidable differences have not prevented agreement on the first two daughter directives for dangerous substances and the United Kingdom hopes that unanimous agreement on their provisions can open the way to progress on the others. If so, this fourth and most recent principle of water quality policy in the United Kingdom will have more influence in the future than it has had so far.

12.6 CONCLUSION

This account of the four principles of UK policy for the protection of the aquatic environment has brought out that even those which are of comparatively long standing have been evolving over a generation and have indeed developed considerably in the last decade. The European commitment has perhaps put them under some strain but is fundamentally compatible with them, and has played a part in their recent development. None of these principles can be absolute (as the House of Lords' Committee said of PPP); but together they should provide a sound basis for the development of policy and practice in the protection of the aquatic environment.

© Crown copyright 1983.

Discussion

Chairman: **H. FISH, OBE**
Chief Executive, Thames Water

There was considerable interest in the chapter by Mr Bjerre. One speaker referred particularly to the problem of sewage sludge disposal at sea and stated that the Oslo and Paris Commissions sometimes gave the impression that they were against all dumping of waste at sea. He was pleased to note that Mr Bjerre's chapter indicated that this was not the case and that they were more concerned with the disposal of toxic materials in the sea. It could be argued that the seas around the UK were a resource allowing sewage sludges to be disposed of safely without causing environmental harm. Mr Bjerre's statement that the Oslo and Paris Commissions were more concerned with the detrimental effect of the toxic constituents of wastes dumped at sea was also welcomed. The sewage sludge disposers are also concerned wih the effects of toxic components.

The need for limits on the concentration of these components in sewage sludge with regard to disposal on land was accepted. It was also accepted that there was a need for limits on the toxic components of sludge disposed at sea relative to the size of the disposal ground in order to prevent environmental decline. The Severn Estuary disposal ground was cited as an example where no environmental damage was being caused by sludge disposal. Evidence was also quoted of local fisherman congregating around sea outfalls because they were a rich fishery area. However, there was a fear that within 10 years there would be moves to prevent the UK from disposing of sewage sludge at sea. Mr Bjerre was asked if he thought this was the intention of the Commissions or were they more concerned with environmental damage within the marine environment.

Mr F. Bjerre answered by stating that the Commission was divided on the question of sewage sludge disposal. He could not foresee the other signatories to the Conventions forcing the UK to comply with a ban on sludge dumping without agreement from the UK. Other countries within the Oslo Convention do have a different view as shown by the fact that Finland, Sweden, Denmark and West Germany have agreed not to dump sewage sludge within the Baltic

Sea, due to the little dispersion afforded by the tides in the area. Mr Bjerre concluded by affirming his personal view that the Commissions would not introduce such a ban whilst a contracting party was so steadfastly against such a course of action.

It was pointed out from the floor that much of the attention of the Oslo Commission was directed towards control of the disposal of sewage sludge. However, when the amounts of contaminants within the sludge were compared with the quantities of the materials entering the seas of the Convention area from other sources it was noted that input of these materials in sewage sludge was small. Riverine inputs of these substances were found to be much greater. The Paris Commission, however, does not devote as much effort to control of this substantial source of contaminants as it does to the control of inputs from sewage sludge. Mr Bjerre was asked whether, in his personal opinion, too much attention was being given to a relatively minor input.

Mr Bjerre replied stating his belief that if the Paris and Oslo Conventions had been combined at the outset a different emphasis in terms of controlling inputs of toxic substances would have been made.

However, the remit of the Oslo Convention only covers dumping at sea. Under the Paris Convention, it is correct to say that the impact of riverine inputs is not being assessed in a quantitative way. The Paris Commission is trying to remedy this imbalance and has recently set up a working group with the specific tasks of assessing the information available on riverine inputs. One of the problems encountered so far has been the difficulty on agreeing a uniform comparable methodology to calculate the size of riverine inputs. The Paris Commission is thus hoping to obtain this kind of data in order to put the input from sludge dumping into context.

Mr A. J. Fairclough noted that although discussion was focused upon disposal of sludge at sea, it was important to realise that sewage sludge was in fact a resource. He reminded the conference of an EC proposal currently before the European Council on the disposal of sewage sludge to agricultural land. One of the objectives of this proposed legislation was to increase, encourage and make possible the re-use of sewage sludge in agriculture. In this respect the UK does not re-use as much of its waste as some of its European counterparts. Mr Fairclough concluded that increased re-use of sewage sludge in agriculture might prove to be a better alternative than disposal at sea.

Continuing this theme of maximising sewage sludge disposal to land a speaker said that it was often the framing of a directive, and particularly the various limits placed on parameters of sludge quality, which determined the sludge disposal strategy of a water authority. Thus the limits on pH within such a directive might preclude the land disposal option in some areas.

The Chairman then asked for questions or comments on related topics other than sewage sludge. Delegates were then shown four slides depicting the loads of various metals carried by UK rivers to the seas around the coast. The loads had been calculated from the data collected from the 250 sites within the Harmonised Monitoring Network. The loads had been determined by simply multiplying the concentration of a given determinand by the flow measurement at the site. Two determinands, cadmium and nitrate were used to illustrate the variation in riverine loading around the country.

Mr Bjerre responded to these slides by stating that the annual estimate of cadmium input to the sea was 7 tonnes compared to the riverine load of cadmium depicted in the slides as approximately 200 tonnes per year. He thought that the Paris Commission would be very interested in this kind of data.

A statement made in Mr Fairclough's chapter on 'the need to modify the Surface Water Directive' was welcomed by one speaker who wondered whether this intention could be extended since by 1985 the Drinking Water Directive would be implemented and thus the need for the earlier Surface Water Directive would be negated, as the Drinking Water Directive could satisfactorily protect the consumer's water quality. Mr Fairclough was asked whether he agreed with this assessment and whether the European Commision had any procedure on the withdrawal of a directive.

With regard to the withdrawal of the Surface Water Directive, **Mr Fairclough** thought that it had a different objective to that of the Drinking Water Directive and thus disagreed with the interpretation given above. On the matter of withdrawal he said there was such a procedure and that it was used frequently, as shown by the changes in proposals for a directive.

Discussion then turned to the apparent conflict between the Environmental Quality Objective (EQO) and Uniform Emission Standard (UES) approaches to the control of discharges containing dangerous substances. The agreed aim of most of the European legislation incorporating these approaches was the protection of the environment. The purpose of the EQO approach was to relate though that it had a different objective to that of the Drinking Water Directive discharge control limits, or Environmental Quality Standards (EQS), to the assimilative capacity of the receiving environment. This was in direct contrast to the UES approach, where the discharge limit was determined by the best practicable method of treating the waste from a given industry. It could be argued that UES limits were a compromise in that they did not relate to best technical or best possible means of treatment but instead related more to an economic assessment of the treatment a given industry could afford. The resulting limit is then promulgated without any requirement that the resultant environmental impact is acceptable. This applied equally well to EC directives, such as regulating the discharge of cadmium to the aquatic environment, where there is no obligation on Member States opting for UES control to monitor the environment to ensure that no damage occurs.

Delegates were shown a table of the UES and EQS values for freshwater from some agreed and proposed 'daughter' directives of the Directive for the Control of Discharge of Dangerous Substances to the Aquatic Environment.

The ratios between these values were also presented and they showed that the available dilution in the environment varied dramatically between directives, the greatest ratio being one of 10^6 taken from a proposed Directive for the Control of Lindane in Discharges to Aquatic Environments. It was stated that the size of these ratios meant that for a small discharger a relatively small volume was required under the UES approach to comply with the environmental quality standard in the directive, whereas a large discharger requires a great amount of dilution. Thus it could be concluded that the control upon the discharger exerted by the UES approach was much more than that enforced under control by an EQO approach.

Mr Fairclough answered these points by saying that although he did not have copies of the relevant directives in front of him, he was quite sure that dischargers could not discharge greater loads of the particular toxic substance by merely diluting their effluent down. He went on to state that he did not wish to join the controversy of the relative merits of the two approaches since his role was simply to enforce the agreed directives, which allowed Member States to control discharges by either method. He acknowledged that under a uniform emission standard approach there was no requirement to monitor the environment and indicated that this might be an area which could be pursued in the future. He continued by observing that there were strong arguments for both approaches and that those in favour of the uniform emission standard approach were overlooked. The UES approach had been put forward in order to control a relatively small number of highly dangerous chemicals, because of their toxicity, persistence and bioaccumulation, about which little was known. Thus the rationale which produced the UES approach was that if there was no real need to discharge such chemicals, then discharge should be avoided or strictly controlled.

Mr Fairclough concluded by explaining the variation in the ratios of UES: EQS in the directives as shown to the delegates was due to variation in the best available technology upon which the UES values were based.

Dr D. G. Miller continued the debate over the EQO versus the UES approach, by explaining that the background to both approaches was understood. He found difficulty, however, in finding a common link between them and asked if the UES approach was applied to industry based on best technical means, then was it not necessary to follow the process through to establish that the UES approach does protect the environment? The scientific community was worried because the efficacy of the UES approach was not being checked by environmental monitoring. This worry increased when policy statements were published, such as in a recent paper by an official of the European Commission, stating that the importance of the UES approach is in relation to unfair competition, whereas if environmental protection is the objective, then the EQO/EQS approach must be used. Dr Miller concluded that the conflicting philosophy of these two approaches needed clarification.

Mr Bjerre commented upon the EQO/UES debate by noting that under the Oslo Convention, the UK and Portugal, who both used an EQO approach had to monitor the effect of this control approach on their marine waters. The same was also true of those countries who aligned with the UES philosophy. However, to date not enough data had been collected to make an objective assessment of the efficacy of the two approaches.

Dr R. F. Packham wished to rectify any negative impression which Mr Fairclough may have gained regarding his own attitude towards EC directives. Most of Dr Packham's criticisms were levelled at the technical annexes of directives concerning water for a particular use. WRC had received requests on many occasions to comment on the technical content of the directives and, as a result, considerable time and effort had been expended by staff of the Centre in compiling these comments. Dr Packham expressed his dismay that in most cases these opinions and comments appeared to have been largely ignored.

Referring to the political aspects involved in the agreement of directives as mentioned in his chapter, Dr Packham accepted that considerations of problems relating to particular large rivers spanning several countries, such as the Rhine, were of general concern. However, in this context he did not understand the rationale behind setting a standard for sodium in drinking water, as had been the case in the EC Drinking Water Directive, when the World Health Organization did not believe that a limit to sodium in drinking water on the basis of health was required.

Mr Fairclough then made the point that it is important that the proposals which the Commission puts forward are based on sound scientific research and are sensible. All Community law was agreed by representatives of the UK Government. Thus it is not the responsibility of the Commission to decide how the views of the WRC or the water authorities should relate to the UK Government and to what influence they should have during the long negotiations which accompany the passage of any directive, which in the end is a political agreement. It is very important to realise that an agreed EC directive is not simply the opinion of the European Commission translated into legislation, but an agreement between all Member States. Thus any dissatisfactions one may have with a given number for a standard in a directive cannot be laid at the door of the Commission. Due to the fact that a directive is agreed by many governments, each with their own interests, it is not surprising that the final legislation contains some element of compromise.

Mr Fairclough was asked why the Commission did not publish the scientific justification for the standards and limit values in EC proposals for directives? An example from the most recent proposal to control discharge of waste from the titanium dioxide industry was given. Although the explanatory memorandum for this proposal stated that it was being put forward 'because of the polluting effect of wastes from the titanium dioxide industry', it did not at any point mention what the polluting effects were, nor did it provide references for the work in which such effects were demonstrated. The conclusion could be drawn that the Commission did not know what the effects were, and furthermore, most explanatory memoranda to EC proposals were often long on platitudes, but very short on facts.

Mr Fairclough explained that the original Treaty of Rome required that all legislation considered by the Council had to incorporate justification of the proposal for that legislation. In general the information upon which proposals are made is available, although not published due to financial constraint. Copies of the studies comissioned are available, which contain the assumptions upon which proposed directives are based. The remainder of the scientific information is obtained by consultation with scientific experts. Thus there is exchange of scientific opinion between national experts before a proposal is put forward.

Mr D. Hammerton (Director, Clyde River Purification Board) provided a written comment on Chapter 12 — The UK Approach — Environmental Quality Standards by Mr J. A. L. Gunn. He writes that the principle of unified management for the water cycle as adopted by England and Wales was proposed by the Royal Commission on Local Government and was debated in Scotland ten years or so ago. The proposal, whereby water pollution control together with sewerage and water supply would be in the hands of nine regional councils was opposed by many bodies. The Royal Commission on Environmental Pollution (Third Report) recommended that the independent river purification boards should be retained and eventually this recommendation prevailed. As a result the river purification boards of Scotland were enlarged and strengthened. Thus Mr Hammerton makes the point that, in the UK we have two different systems of managing the aquatic environment, each being appropriate to the needs of its respective area. Furthermore, it is Mr Hammerton's personal opinion that full development of the EQO/EQS system was not dependent on the creation of regional water authorities and there is no evidence that river water quality managed under one system is any better than under the other.

As a final point on the origins of the EQO/EQS approach, Mr Hammerton recalls the work of the Royal Commission on Sewage Disposal which some seventy years ago proposed standards which were based on a thorough form of environmental impact assessment and designed with clearly defined river water quality objectives in mind.

Referring to the anticipation of environmental effects in Professor Salzwedel's paper, it was recalled that at a recent meeting of the Royal Society, the deposition of sulphur and nitrogen compounds was quoted as being at a rate of 50 kg/ha which was approximately split half each between the two compounds. It has been estimated that 90% of this deposition falls on the UK yet our forests do not appear to be affected. The anticipation work in Germany had considered the impact of deposition of sulphur compounds, but had tended to overlook the importance of nitrogen compounds, which were produced as a by-product of unburnt hydrocarbons from motor vehicles. Thus no clear cause and effect relationship between aerial emissions and environmental damage had been shown. The anticipation approach needed to contemplate a few more of the complex sources of aerial pollutants and much more scientific research was required.

Professor Salzwedel responded by informing the conference that most of his work was concerned with photo-oxidation problems. As nitrogen compounds are produced during photo-oxidation this seems to be an important pathway. The main problem is to try to influence the motor industry to change the design of its car engines. It is clear that there is no direct link between the products of photo-oxidation nor deposition of sulphur compounds and the environmental damage observed. Thus it is a question of anticipation of the detrimental environmental effects and taking action to prevent them.

Professor Salzwedel was asked whether the Lander actually used the Federal Emission Standards mentioned in his talk, or whether the Lander regard them as a maximum licensing level and calculate the actual licences on the basis of the existing water quality in the rivers.

Professor Salzwedel confirmed that, on the whole, the Lander enforce the minimum requirements. However, in some difficult situations due to problems of water quality, such as the Neckar, the general practice is to go far beyond the minimum, with consequent economic effects. In general it is difficult to assess the actual situation because there is a great deal of conflict between the Federal Government and the Lander upon these matters.

Part 4

OPERATIONAL ASPECTS

CHAPTER 13

Pollution control in practice

B. ALEXANDER and E. HARPER
North West Water Authority

13.1 INTRODUCTION

The Conference is concerned with many philosophical questions, and detailed scientific examination of environmental protection issues. This chapter deals with the practice of pollution control and is therefore concerned with how to achieve results in the field. The old saying 'practical politics is to do what you can, not what you ought' may not have been written about pollution control, but it could have been.

There are probably no cases of existing pollution where the remedy does not involve some capital expenditure and where gross pollution occurs the costs in capital and revenue can be very substantial. All effective pollution control and pollution prevention measures necessitate careful planning of capital expenditure, good operation of the facilities provided by this investment, and monitoring of the effectiveness of the accepted solutions. None of these aspects will of themselves guarantee success as they are interdependent.

This chapter deals mainly with controls as they relate to river pollution where issues such as recreation and amenity are important. A similar practical approach is applied to other areas of pollution control but there is much less room for subjective judgement where public health is concerned, for example measures specifically to protect potable water in supply. It should be remembered that what may be yesterday's pollution problem in one part of the region is potentially tomorrow's in another part and care must be taken not to convert a problem of river pollution into one of land or air pollution.

13.2 ELEMENTS OF SUCCESSFUL POLLUTION CONTROL: SOME PLANNING ASPECTS

The aims of pollution control are clearly, if imprecisely, set out in legislation. Rivers and other inland or coastal waters must be maintained or restored to

[Ch. 13] **Pollution Control in Practice** 199

wholesomeness, and the supply of potable water to consumers must be wholesome. Some see the lack of precision in the legislation as a disadvantage but in practice it allows experience and scientific judgement to be applied in its application to particular problems. Decision makers rarely deal with problems which are precisely defined and practicable solutions to real problems inevitably must resolve conflicting interests. Greater precision becomes important when action is taken to define a target which at least is in the direction of the aim so that success can be measured by which part of the target is hit. Logically there are four steps to be taken to solve effectively a problem in pollution control.

13.2.1 Step 1: Identification

There is usually ready agreement whether or not pollution exists, what is often contentious is a more precise definition of what constitutes the problem. Such definition tends to be coloured by an awareness of the next steps and the approach of North West Water Authority (NWWA) is no exception. Anyone looking over a bridge on the River Mersey during the last century would agree that there is a problem, but what exactly is it? At this point it is pertinent to consider a definition of pollution and the following has the advantage of being an accepted European view. 'Pollution' means the discharge by man, directly or indirectly, of substances or energy into the aquatic environment, the results of which are such as to cause hazards to human health, harm to living resources and to aquatic ecosystems, damage to amenities or interference with other legitimate uses of water [1]. The Control of Pollution Act, 1974, also puts specific emphasis on injurious pollution and in relation to a stream defines it as pollution injurious to the flora and the fauna of that stream. The most significant factor in practice of these definitions, and all generally accepted definitions of pollution, is the element of harm, or injury, or a potential to cause hazard to man. These definitions clearly imply that there are levels of potential pollutants which are acceptable in the environment without causing pollution.

The region covered by NWWA has only some 10% of the total land area of England and Wales, and contains only 14% of the total population, but has 36% of 'grossly polluted' rivers as defined by the National Water Council classification system and published data in 1980 [2]. Regrettably this is a greater polluted length than occurs in the region of any other water authority. The classification system contains quality criteria which are defined in reasonably precise terms, linked with more subjective elements and references to current and potential uses. It has therefore been adopted by NWWA as an aid to defining more precisely the nature of the problem. In 1976 the Authority adopted a policy for river water quality which has a long term aim of restoration and protection, and a short term aim to prevent deterioration and as far as capital is available and local needs exist, to improve river water quality. A fuller statement including the 'practical steps towards the short-term aim' is included as Appendix A (section 13.8). This was supplemented in 1978 by a consultation document [3] on long-term river water quality objectives which was circulated widely throughout the North West. The proposals received general acceptance and were:

(i) Existing Class 4 and Class 3 rivers to be improved to at least Class 2, where practicable.†
(ii) Existing Class 2 to be improved to Class 1A or 1B where reasonably practicable.
(iii) Existing Class 2, where it is not reasonably practicable for improvement because of diffuse sources of pollution, to be maintained in their present classification.
(iv) Existing Class 1A and Class 1B rivers should be maintained in their present classification.

Attainment of the long-term objectives (better perhaps thought of as aims as a time limit was not included) will place 70% of the total length of inland rivers and canals in Class 1A or 1B, and 30% in Class 2. It was estimated then that some 380 km of the 5,597 km considered would not achieve Class 2 in what was seen as a realistic assessment of foreseeable availability of finance and technical resolution. About two-thirds of the length is affected by storm sewage overflows, and the remainder by intractable industrial wastes.

It can be seen therefore how the aim to remedy pollution has conditioned the way the NWWA has defined its pollution problem. If the present aims are achieved then the rivers will be a great deal cleaner and there will be much greater potential for multiple use but they will not be pure. It will always be possible to set higher aims in the future and perhaps eventually esoteric discussion of the exact meaning of wholesomeness will be relevant to practical achievement of results.

13.2.2 Step 2: Options

When existing problems have been recognised and defined then consideration can be given to the proposed solutions. The possible effects of alternative solutions must be evaluated as it is possible to replace one problem with a different one, perhaps harder to resolve. If the definition given earlier of pollution is adopted then clearly all new schemes must prevent pollution not control it. This does not mean that preservation of water quality at all costs is the aim, rather conservation and full use of the adaptability of natural systems. Such considerations inevitably lead to decisions, implicit or explicit, on safety margins. In practice it would be ludicrous to spend more money on assessing the risks of potential pollution rather than spending the same sums on preventing the risks

---

† This category includes rivers which will not attain Class 2 unless something is done to solve the problem of storm sewage overflows or problems of intractable waste from industry.

altogether. On the other hand for major schemes such as water resource studies an environmental appraisal of not only water quality pollution but all the environmental issues may be considered necessary. NWWA employed specialist consultants to carry out such appraisals as part of its examination of options for the Public Inquiry at Ennerdale Water and when considering regional water resources [4] until the end of the century. At the other extreme the construction of a small sewage treatment plant treating domestic wastes and discharging into a major clean river requires more than a subjective judgement. This will be conditioned by what is reasonably cost-effective in sewage treatment. On clean rivers where possible pollution effects are subtle and consequently difficult to predict and measure subsequently as the whole aquatic ecosystem needs to be considered, there is some justification for erring on the side of conservatism, or preservation, as a cost effective and practical approach.

It is at this step that a precise definition of the problem becomes valuable as it will often reduce the options to be considered; many can be eliminated as providing solutions which go beyond, or are insufficient, to match closely the properly identified problem. If the problem was considered obvious enough not to need precise quantitative definition then it must be accepted that only the same subjective approach can be applied to the consideration of options, and their effectiveness will never be capable of being measured in other than relative and approximate terms. Careful consideration now can produce solutions which not only make immediate and substantial improvements in pollution control, but by application of good design can significantly reduce the potential risk of future pollution.

13.2.3 Step 3: Preferred solution

A preferred solution in pollution control terms should clearly solve the problem, but what does that mean in practice? If a solution is 100% certain of achieving the elimination of a problem it probably means that it is over-designed, over-expensive and by utilising scarce resources, preventing another problem from being tackled. Some assessment of the degree of risk in not achieving a 100% solution, conditioned by a reference to the consequences of an error, is usually evaluated subjectively, often subconsciously, as part of the exercise of judgement. Effective pollution control for the future demands that this step be made more conscious and that a deliberate effort is made to quantify the risks of failure, or more positively the probability of success, and quantitatively assess the consequences of error and the anticipated failure rate. The ideal solution will optimise the use of capital/revenue in sewerage and treatment facilities (private and public) with the uses and potential uses of the stream.

The water industry is already a capital intensive industry with a ratio of capital to revenue of 68% compared with the gas and electric industries' investment of 25% and 18% respectively. The first ever national strike in the water industry in 1983 highlighted many areas still remaining where existing proved systems could have provided a continuing service for a longer period without manual intervention. Pressure on revenue expenditure by the tightening of performance aims by government will reinforce the existing trend towards capital intensive solutions. In pollution control terms this must be welcomed as with application and appraisal of techniques to increase systems reliability it should be possible to design treatment plants to meet their 24 hour per day duties without unacceptable failures. Experience proves that many pollution incidents stem from human failures deliberate or otherwise, and that plant failure even with the present low level of application of process control technology causes relatively few incidents. It must be stressed again however that capital investment cannot completely replace the knowledge and skill of a human operator and in many cases still the most cost effective solutions to existing problems necessitate increased man-power, chemical costs and energy, which can conflict with performance aims. Even automated plant requires maintenance.

13.2.4 Step 4: Implementation

This step is taken to cover design, construction, commissioning and operation and is the step when concepts become a reality, any weaknesses in the earlier steps will become glaringly apparent. It is, however, too late to change as a designer must not be expected to decide what a scheme should achieve, nor must a designer hinder the preferred solution because he has a scheme which will provide the solution to the problem if only it were redefined. This should be the stage by which a clear performance specification has been derived, the task now is to provide the least cost solution to meet this specification without additional safety margins. If performance specifications have not been set no amount of post-completion monitoring would provide the information to prove the effectiveness in pollution control terms of the scheme. At the very least the failure of a scheme to solve a problem should result in valuable lessons to prevent a recurrence. Good operation is of course vital to achieve the required level of pollution control at least cost and is largely taken for granted in this paper. It cannot be over stressed that an apparent least cost solution in capital or revenue terms which does not achieve its aims is not cost effective, although it may be relatively cheap.

Paradoxically perhaps, as major pollutions are reduced by capital expenditure the need for greater reliability in plant operation increases. Rivers respond to reduced pollution by supporting a richer and more varied life which is more sensitive to pollution. A treatment plant which produces bad effluent occasionally may be acceptable in a Class 4 river, it may even be acceptable in a Class 1A river which has an adequate safety margin of natural self-purification, but it will almost certainly have a devastating effect on a marginal Class 2 river where fish are perhaps surviving though not thriving.

13.3 SOME ENFORCEMENT ASPECTS

Other authors are discussing the scientific criteria of setting standards, and how compliance should be measured, but what should be done if they are not achieved? At best the expected safety margins will be eroded, at worst serious pollution will occur, control will have been lost. An immediate reaction is to refer to the legal penalties under the various Rivers (Prevention of Pollution) Acts. In a Magistrates Court the maximum fine is £1000 and on indictment an unlimited fine is possible. For a continuing offence it is also possible for the Crown Court to impose imprisonment of up to 3 months and/or an additional fine of up to £10 per day with a £1000 maximum. In practice the average fine in 1982/83 imposed in the Magistrates Courts in the North West was only £212.5, and the single case heard in the Crown Court resulted in a fine of £1000. This compares with prosecutions taken by the Manchester Ship Canal Company under the Prevention of Oil Pollution Act 1971 applicable to navigable waters where the maximum fine is £50 000 and fines up to several £1000's have been imposed. Oil pollution incidents under the other Acts result in fines of typically £100 to £200. However, it must not be overlooked that companies also have common law liabilities and certainly there have been several cases where companies have agreed to restock fishing rivers which have been affected incurring costs of £10 000 to £20 000. Similarly although the typical fine for an oil pollution incident is low, companies are expected to foot the bill for the clean-up operation which follows, and can be faced in a major incident with costs of £20 000 to £30 000. The Control of Pollution Act 1974 will give the water authority a new duty to remedy pollution which damages flora and fauna, together with the power to recharge companies. It seems likely that this new duty will prove to be a greater financial deterrent than fines.

Fortunately the climate of public opinion is strong enough, and reputable companies sensitive enough to such public pressure, to ensure that financial penalties are not the prime motivation in abiding by the law. This corporate responsibility is perhaps the strongest single factor contributing towards effective pollution control, and in practice control strategies and tactics must build on this factor. The fallibility of the human element as mentioned earlier is often attributable to ignorance of the consequences of actions, and occasionally sheer unthinking negligence, rarely does it appear as deliberate actions after an assessment of the pollution consequences or indeed the risks of prosecution. A good example is illustrated by the many cases of oil pollution which stem from holes which have been constructed (often merely knocked) in the bottom of an oil bund. It appears that a significant number of plant operators never think beyond the fact that water collects from rainfall in the bund more often than oil spillages occur; they have presumably carried out their own risk assessment. When the exception does occur the company involved is always genuinely surprised and expresses sincere regret, the oil which is lost may well have cost several thousand pounds notwithstanding the consequences.

13.4 PREVENTION BETTER THAN CURE

Any discussion of penalties and costs already accepts failures occur and that pollution will result. The best pollution control is centred on prevention and limitation of actions so that, even if contamination occurs it is not at a level which causes damage. It is here that the local officer is so vital be he a river inspector, trade effluent officer or fittings inspector. Such officers can often spot potentially polluting events, such as a hole knocked in the bund of an oil tank, or a cross-connection which could contaminate potable supplies, and bring it to management's attention before it is too late. A general education role creating an awareness of pollution effects must help to reduce pollution incidents, it has certainly increased reports of pollution incidents from the public. Effective public information including films, talks, publicity leaflets, etc. are an important part of pollution prevention. A visit from an inspector is always salutary in reminding potential polluters of their obligations to society. This role is often considered analogous to the policemen on the beat perhaps supplemented by the crime prevention officer; but this is not enough, polluting discharges are often intermittent and pose detection problems which would tax Sherlock Holmes.

Urban areas present special problems, many watercourses are culverted, and most towns in the North West are drained on a combined system with many overflows constructed to relieve an overloaded foul system to the nearest watercourse. When a major polluting discharge to a river is improved invariably a multiplicity of smaller discharges previously considered relatively insignificant begin to show their effects, and can prevent the full benefit of the major scheme being achieved. These problems are associated particularly with sewerage systems constructed last century. In the proceedings of the Royal Commission on Sewage Disposal [6] in 1905 evidence was even then being given that there were 215 storm sewage overflows on the Manchester system to the Davyhulme works when only 37.5 of the population used water closets, the remainder were on pails and middens. The overflows were intended to operate when the sewage was diluted 5:1. It is probable that many of these overflows still exist when WCs are practically universal. NWWA seems to have coined the phrase 'underground dereliction' [7] to describe the state of the 'assets' which it inherited when assuming responsibility for sewerage in 1974; some pollution is inevitable. Major capital expenditure is the only pollution control measure which will be effective in the long-term. In the meanwhile constant vigilance in observing the effects of overflows as these are often aggravated by sewer blockages, which can be alleviated by speedy palliative action, and heavy maintenance costs are the price which must be paid. Members of the public are valuable for drawing attention to incidents which necessitate investigation and in fishing rivers can be relied upon to make their views known on any cases of pollution.

Many old industrial sites have a labyrinth of drains and after closure of the original industry, which was in the North West often textile or paper mills, have now evolved into multi-use sites which makes tracing a source of an intermittent polluting discharge difficult. Even modern purpose-built industrial estates often have cross-connections between foul and surface water drains, and also contaminated areas draining to watercourses. It is important to establish links with planning authorities as this provides one means of seeking to prevent pollution by consideration of potential problems early in any 'change of use' where planning permission is needed.

There has been in recent years heavy emphasis on the United Kingdom approach which is based on environmental quality objectives, NWWA fully supports the general view. However, again stressing that in practice there is much subjective judgement to be exercised, the Authority when formulating its river water quality policy (summarised as Appendix A – section 13.8) also qualified its approach to include two points to have particular emphasis:

(i) the policy should not be rigidly adhered to if this would not be consistent with the most efficient use of the Authority's limited capital resources;
(ii) as a general practice in both the long and short term all practical, cost effective, means shall be taken to minimise or eliminate potentially polluting discharges.

Encouragement of a positive attitude to prevent pollution is a necessary part of pollution control. The emphasis in this river water quality policy is to ensure that a presumption that the river can be considered as a convenient recipient for minor discharges does not lead to adverse additive effects. A pragmatic approach applying some constraint can be justified as reasonable in the saving of scarce resources which would be needed for a precise evaluation of every potential problem. It should be remembered that in an appeal to the Minister under the Rivers (Prevention of Pollution) Acts the test is whether a consent has been unreasonably withheld, or that the terms of any consent are unreasonable. In the terms discussed earlier under 'Some Planning Aspects' the options must be considered with environmental quality objectives in mind but without any overriding presumption that the river is a convenient dustbin which can be used until its capacity is taken up. The size of the dustbin is not after all exactly known and it is prudent to have some spare place for an unexpected load of rubbish!

13.5 MONITORING THE EFFECTIVENESS OF POLLUTION CONTROL

It has already been said that if pollution occurs then controls have failed. Routine chemical monitoring can help to detect contamination which if continued could result in pollution. In NWWA 39 key sampling points on the major rivers and tributaries are regularly monitored and are one element of the river water quality policy (See Appendix A(iii) – section 13.8). This monitoring is intended to show only general trends and is not particularly sensitive to changes at individual premises. Nonetheless this monitoring has occasionally detected significant changes due to particular discharges. As an example one year it was necessary to downgrade a river from Class 2 to Class 3 due to increased ammonia levels. The source was tracked back to a change in operational practice at a particular sewage works. The change had reduced nitrification processes in order to save energy costs. This had not significantly changed carbonaceous oxidation and only BOD and suspended solids limits were included in the consent. This illustrates that even water authority process managers do not always consider the full consequences of their actions, and that consents do not always include every constituent which may affect a river.

The success of the practical steps which have been taken to control pollution are well demonstrated by the graph at Appendix B (section 13.9). This shows the biochemical oxygen demand and dissolved oxygen levels of river water at Howley Weir, Warrington, which is the freshwater limit of the River Mersey before its tidal reaches. The graph also shows the value of the key sampling point approach as a means of showing trends in river water quality. The differences from year-to-year are partly due to meteorological factors and are not due to pollution control measures alone which cannot therefore be isolated and expressed statistically but the trend is clear. It may be of interest that the temporary effects of the industrial action in this catchment in February 1983 was to produce river conditions which were commonplace 20 years ago.

This general monitoring cannot prevent pollution and biological monitoring is a better tool than chemical monitoring for seeking to achieve this end. Changes in aquatic communities, especially in the cleaner rivers, can detect changes which if continued would constitute pollution. Regular surveys show the continuing health of a river and thus enable resources to be devoted to catchments where changes are occurring. They also show the effects of intermittent pollution for a longer period than the often transient chemical discharge, even if it is not possible to pinpoint the source of pollution. Biological monitoring is an essential complement to chemical monitoring for pollution control purposes.

13.6 CONCLUSIONS

Pollution control must become an integral part of the consideration of virtually all activities within a catchment if it is to be effective. Remedial action to control levels of pollution if left to the final operational stage are often more difficult, more costly, and on occasion impossible to implement without starting again. To end on a philosophical note to a practical paper, the prayer of Reinhold Niebuhr is considered appropriate.

> God, give us grace to accept with serenity the things that cannot be changed, courage to change the things which should be changed, and the wisdom to distinguish the one from the other.

13.7 REFERENCES

[1] European Communities (1976) *Council Directive on Pollution Caused by Certain Dangerous Substances Discharged into the Aquatic Environment of the Community* (76/464/EEC) May.
[2] National Water Council (1981) *River Quality; The 1980 Survey and Future Outlook.* December.
[3] North West Water Authority (1978) *Inland River Water Quality;* Consultation paper on suggested long term objectives. November.
[4] North West Water Authority (1978) *Environmental Appraisal of Four Alternative Water Resources Schemes, Haweswater, Borrow Beck, Morecambe Bay, Hellifield.* November.
[5] National Water Council (1976) *Paying for Water.* April.
[6] Royal Commission on Sewage Disposal (1905) 5th Report, Appendix I Evidence of Gilbert J. Fowler, Consulting Chemist to Manchester Rivers Department, May.
[7] North West Water Authority (1978) *Underground Dereliction in the North West.* J. G. Lloyd, Royal Society of Arts Seminar, February.

13.8 APPENDIX A − NORTH WEST WATER AUTHORITY − POLICY FOR RIVER WATER QUALITY

Long-Term Aim
The restoration and protection of the river water quality of the region.

Short-Term Aim
First to prevent deterioration of the present situation and second, as far as capital is available and local needs exist, to improve river water quality.

Practical Steps Towards Short-Term Aim
  (i) To protect, and as far only as is necessary, to improve the rivers which provide, or are expected to provide in the foreseeable future, the high quality potable water supplies of the region.
 (ii) To prevent deterioration of river water quality that would result in an unacceptable change of character of a river, although some local deterioration of water quality in individual river stretches might have to be accepted within this policy.
(iii) To prevent deterioration of river water quality at certain specified points in river basins which are of special significance.
(iv) To reduce and, as far as possible, to eliminate gross pollution of rivers that constitutes a nuisance.
 (v) Where there are pressing local needs, to take advantage of any opportunities for improving river water quality that might arise.

The relative priorities of the steps should be treated as being generally in accordance with the order in which they appear above.

13.9 APPENDIX B – RIVER MERSEY: TIDAL LIMIT

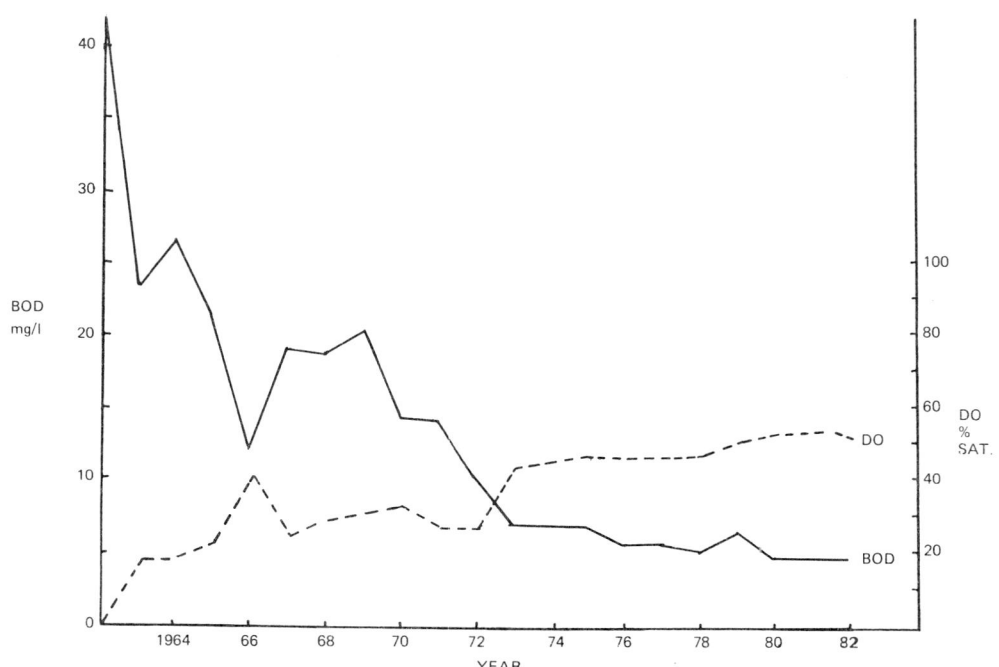

CHAPTER 14

Spillages – an operational approach

J. A. YOUNG and R. J. WHITAKER
Wessex Water Authority

14.1 ABSTRACT

In this chapter the authors have set out the problem arising from the manufacture, transportation, storage, use and disposal of a wide range of oil and chemical products and go on to describe the interest and involvement in this subject of the predecessor river and water boards, water companies, river authorities and those local authorities involved in sewage and sewage disposal.

The authors go on to show how the industry both before and since re-organisation in 1974 have developed their response to incidents involving spillages of oil and chemical products.

The chapter highlights some of the earlier problems with other authorities with interest in spillages. The primary emergency services, for example the Police and Fire Brigades, and in particular the problems caused by the Hazchem code and its sometimes conflicting advice to those first on the scene and possible subsequent damage to the environment.

The authors go on to describe how their own particular water authority deals with such reported spillages from its Regional Operations Centre and of the manpower and equipment used to deal with such spillages.

14.2 INTRODUCTION

The ever increasing manufacture, transportation, storage, use and disposal of a wide range of oil and chemical products have made it inevitable that spillages of these oils and chemicals will occur and will continue to increase. Such spillages are either a danger or a nuisance to the population and to the environment. Such spillages can be gaseous, liquid, semi-liquid, viscous, solid or powdery. Spillages can also result in the materials being spilt reacting with other liquids and chemicals, or the air itself. This chapter limits its comments to the effect of such spillages on the hydrological cycle both the natural cycle and the man-made loop in that cycle of abstraction of water, its treatment, distribution, use and return through the sewage treatment system into rivers or the sea. As the chapter describes an operational approach, it also covers the area of gaseous spillages inland and coastal with particular reference to their effects on water authority water quality staff as well as the primary emergency services dealing with such an incident.

The water industry's interest in spillages has been a long one. Primarily the involvement was with the river boards and their succeeding river authorities with their pollution control legislation and with water supply undertakings (boards, municipal or companies) anxious to protect their intakes. The operation of plans, manpower and equipment to deal with such emergencies has been developing since the early 1950s and is a continuing process. Prior to re-organisation most river authorities were equipped to deal with inland spillages that occurred from either mobile or static incidents and had for a number of years held discusions with the primary emergency services regarding the satisfactory resolution of hazardous spillages. During this period many problems had to be overcome.

The introduction and use of the Hazchem Code, though a step forward, brought with it a number of problems to the water industry. The Hazchem Code is designed mainly for the Fire Brigade as the primary emergency service for first aid action during the first fifteen minues or so of a spillage. It is quite rightly pre-occupied with the safety of the fireman, reduction of fire or explosion risk and the saving of human life.

The Hazchem Code is an ACTION CODE for firemen on the site of the incident. It tells in very simple terms the basic steps that they should take to protect themselves, protect the general public and the most suitable method of dealing with the spill during its initial impact. This action is expected to be modified as additional information is obtained.

A major step forward was gradually agreed nationally whereby it was recognised by the water industry that at such an incident the primary task of the emergency services is to save human life. Secondly, to reduce any fire or explosion hazard. This has been accepted and it has been widely recognised that if during the course of either of these two operations toxic, obnoxious or other material causes pollution or other operational problems for the regional water authorities then this is a matter which will be dealt with as a cleaning-up operation. After it is decided that there is no danger to human life or risk of a fire or explosion hazard, it is now generally agreed that the competent qualified Water Quality Officer of the regional water authorities sent to the scene of such incidents, will in conjunction with other interested parties, for example, the water authority's scientific staff, the County Scientific Officer or Environmental Health Officers, decide on the best way to neutralise or otherwise make safe or remove toxic, polluting or obnoxious matter. This has resulted in much closer co-operation between the Fire Brigades and the regional water authorities.

Ch. 14]   Spillages – An Operational Approach

A valuable system is Hazfile, a computerised filing system enabling chemicals to be identified by fragmentary evidence, that is a scrap of a label, this may be all that is available after a tanker fire! It also gives additional information on the chemical properties of the spilled material, but does not give guidance to the fire service for the type of action they should take.

Agreement has been reached with the Police over assisting in the access of water personnel to sites of such incidents and Police car escorts have been provided on occasions to motorway incidents. In addition the Authority has reached agreement with the County Councils regarding responsibility for dealing with oil and chemical pollutants whether inland or coastal and a great deal of co-operation exists between the authorities in the use of plant equipment and manpower. Internally, the authority's own divisions are equipped with either landrovers, trailers, or specialised vehicles to deal with a wide variety of spillages.

The faults of the Hazchem Code from the water industry's viewpoint are:

(a) Too rigid, that is fails to distinguish between the action that could be taken to deal with a spill in an area where no surface or ground waters are at risk and the action that could be taken if the same chemical is spilled adjacent to a drinking supply source/catchment.

(b) That although it is an Action Code to be used for the first fifteen minutes or so, the action taken by the Fire Brigade during this period more often than not dictates the action for the rest of the incident.

(c) That fire service believes that the code gives them the right to continue the action already started and not amend their action in the light of subsequent information.

It should be recognised that the action code is arrived at using the best possible information available and where the material being coded is unusual, immediate and long-term toxicity data may be incomplete.

As far as notification of incidents is concerned, all of the incoming calls to the Authority regarding spillages from the primary emergency services come during the entire 24-hour period to the Authority's Regional Operations Centre.

Similarly, any calls from members of the public outside of normal working hours are received at the Regional Operations Centre. During normal working hours call are received also at the Authority's three divisional control rooms and the internal emergency planning system caters for the routing of those calls to Region and to the appropriate duty staff to deal with such incidents. In addition, at the Regional Operations Centre the Authority has a place name gazetteer on a microprocessor system which within a few seconds can pinpoint the location of every village and hamlet within the Authority's area and more importantly to identify the duty staff for that particular day, these being automatically updated by the computer. In addition this system lists the appropriate Fire, Police and Ambulance authority. Typical printouts are shown in Figs. 14.1 and 14.2. Currently, the Authority is developing motorway drainage plans into the

Operational Aspects [Pt. 4

```
Enter Command and press return:  GBATH
1  BATH D.C.              BATH
2. WANSDYKE D.C.          BATHAMPTON
3. TAUNTON DEANE D.C.     BATHEASTON
4. WANSDYKE D.C.          BATHEASTON
5. WANSDYKE D.C.          BATHFORD
6. TAUNTON DEANE D.C.     BATHPOOL
Enter Number and press return:- 1
GAZETTEER INFORMATION
Name:- BATH               Population (1971):-   84670
District Council:- Bath D.C.   Households (1971):-  30225
Grid reference:- ST745645   Area:- 28.72 sq.km.
Microfilm:- 91          ** Kidney Patient - Telemetry B135/A130/S131 **
NFU Contact:- Taunton Office (0823)84408
Cnty Sec:- A Gibson (Hemyock) 680589.  Asst. Cnty Sec:- W Leach (Wellingon) 4617
Area responsible for each function:-
Supply     -  Bath  Stan Hilleard R/P 8093 667 333 92-833068   AB20
Recovery   -  Bath
Sewerage   -  Bath D.C.        Day   :  Bath 92-28411
                               Night :  Bristol 791834
Rivers     -  Bath
Police     -  Avon & Somerset E Division 92-63451     HQ Direct 290721
Fire       -  Avon            Direct Line - 22061     B Division 92-66267
```

Fig. 14.1.

```
Enter Command and press return:- DBS
Bristol Avon Division standby rota (Supply) - Monday 25/7/83 14:08 hrs
Operations Officer          Gareth Jones     R/P 0893 600 120 97-52665    AD19
Pollution Officer           Steve Woods      R/P 0893 667 328 93-261163   AD92
Hydrologist                 Ken Tatem        R/P 0893 600 564 92-859338   AD28
Works Officer               Francis Gay      R/P 0893 667 326 92-314062   AB73
Supply - Chippenham/Devizes Richard Portlock R/P 0893 667 332 92-704562   AA43
Supply - Trowbdg/Warmstr    Alan Ford        R/P 0893 667 330 963-5243    AA82
Supply - Bath               Stan Hilleard    R/P 0893 667 333 92-833064   AB20
P.S. & Res. - Chippenham    Sam Sherratt     R/P 0893 667 342 0380-2757   AA14
P.S. & Res. - Bath          John Chislett    R/P 0893 667 345 92-310096   AB08
M & E (Spt.) - Bath         Clive Hulbert    R/P 0893 667 341 92-28764    AD39
Instruments                 Garry Cox        R/P 0893 667 500 92-313342   AD16
Telemetry                   Robert Eggbeer   R/P 0893 605 815             AD12
Equinox (R/P 0893 667 121)  Chris Kneeshaw   R/P 0893 667 121 888-2585
Enter Command and press return:-
```

Fig. 14.2.

same microprocessor system. When notified of an incident by the Police or Fire Brigade the duty controller obtains the necessary information by keying into the microprocessor the motorway number, its carriageway direction that is north, south, west, east and the marker number. The microprocessor system will then give a grid reference, the river or stream to which that section of the motorway drains and a list of the licensed abstractors and landowners for the first mile downstream. The system also highlights any particularly sensitive areas such as water intakes, fish farms etc. Secondly, the Authority has developed a list of some 1000 chemicals, including those commonly involved in spillages and for which the Authority's scientific staff have a particular interest. The system differs from the Hazchem and Hazfile system in that it is specifically aimed at the effects of such spillages on water in the environment. Especially dangerous or toxic chemicals are highlighted and are flagged in such a way that the immediate attention of scientific staff is drawn to their involvement in such a spillage.

The Authority's well equipped laboratory at Saltford, near Bristol, is capable of analysing a large range of different chemicals very quickly by the use of a gas chromatograph connected to a mass spectrometer.

14.3 TYPES OF SPILLAGES – INLAND
14.3.1 Mobile
As a result of road traffic accidents involving transported oil and chemicals either in tanker vehicles or as packaged or containers, for example, carboys, drums, bottles. This type of spillage would include primary loads being delivered from sites of manufacture or between sites or even in transit between countries. The large number of mixed loads of oil or chemicals to be disposed of are particularly hazardous including those of a caustic or explosive nature and the reaction of some of these loads causing overheating, leaking, corrosion of valves etc.

14.3.2 Static
Normally from bulk or drum storage, pipework, insufficient or non-existent bunding of oil tanks, bulk oil storage for housing estates, refining industry, distribution depots, accidents at factories or process plants and deliberate spillages or illegal dumping.

14.3.3 Coastal and marine
(a) *Coastal*
Oil coming ashore from a shipwreck or deliberate discharging of oil, lost loads, deck cargo drums washed overboard and then ashore, dumped chemicals breaking free and being washed ashore.

(b) *Marine*

Shipwreck or deliberate discharging are the two most common occurrences. The regional water authorities are not responsible for clean-up operations on beaches which is the responsibility of the county councils and the County Oil Officer, but are responsible for those parts of the regional water authority areas which extend into coastal areas, for example, harbours, tidal rivers and estuaries. Operationally, Wessex Water Authority receives details of oil at sea via the County Councils using the Manchgrid system. This system has a numbered grid extending from Southern Ireland, the south west approaches, and the south coast of Great Britain as well as the northern French coast. Any oil spillage occurring within this area is plotted on the grid and the progress of its movement is reported at intervals by shipboard and aeroplane observation, and the numerical grid figures involved in the spread of such oil are reported at regular intervals by teleprinter to the county councils and other interested bodies. The water authority keeps in touch with the County Councils, the Coastguard and Harbour Authorities and is involved in such incidents if the Authority's functions are involved and assist the county council if requested, for example, with the loan of booms or other equipment. Similarly, the Authority provides scientific advice to the county councils where requested on the identification of chemicals or oils in drums or containers washed ashore and are involved in the decisions involving the disposal of chemicals or oil from beaches or coastal areas with the appropriate authorities.

14.4 DEALING WITH AN INCIDENT

Once the Regional Operations Centre is made aware that a spillage has occurred the appropriate Divisional Water Quality Officer is despatched to the site to reconnoitre the situation and to report back. On arrival at the site he will identify himself to the senior police or fire officer at the scene and will liaise immediately with any other officer from the county or district council who may have been called in to assist.

Occasionally, in the case of oils and chemicals some involvement from the oil or chemical industry itself is made available and representatives from those industries may be on site and will certainly be consulted. On one occasion after drums spilled from a lorry the industry representative assured us there was no danger and then arrived shortly after on site by helicopter! Once the primary emergency services have confirmed that there is no danger to human life and that there is no fire or explosion hazard, the duty water quality officer with the fire authority and other interested parties will decide upon the best way to deal with the situation.

In some cases this might be effected by the washing away with copious amounts of water. In others containment and subsequent removal is the answer. The possibility of transferring material from the vehicle involved in a spillage

Ch. 14]　　　　　Spillages – An Operational Approach　　　　　215

to another tanker is also considered, and in addition either neutralising or other form of making safe is used. There are still, however, a number of areas of concern mainly involved in the personnel dealing with the incident. Almost invariably spillages involved in road traffic accidents result in the Police being first on the scene and in almost every case that has been investigated the initial response by the Police has resulted in Police Officers approaching such a road traffic accident or a static spillage completely unprotected from the effects of any gaseous or liquid spillage. Very few constabularies as a matter of course carry in motorway or other Police patrol cars, protective clothing and/or breathing sets. It is not until the Fire Brigade arrives on the scene that such equipment is available. Similarly, it is most unlikely that approaching fire vehicles are given any indication of the wind direction and in the event of a motorway accident normally approach from the same carriageway direction of the site of an incident.

14.5 PLANNING

To ensure that the results of spillages and accidents involving the water authority concerned are dealt with efficiently, it is vitally important that a number of components are fully understood.

The first is that there should be a realistic written Emergency Plan for dealing with a wide variety of emergency situations. The part of the plan dealing with pollution incidents should be framed in such a way that as many as possible of the different types of incidents are covered, and that the appropriate staff are identified and trained and are fully conversant with their role in any particular incident. It is important that the Water Officer attending the scene should identify himself to the officer in charge of the incident and to this end waistcoat jackets marked 'Water Incident Officer' are provided for staff. It is very necessary that this officer stays at the scene to give advice and if he requires assistance from other water authority personnel that these are called to the scene. Some form of identification on private cars is absolutely necessary particularly where Police assistance may be required to transport or guide such vehicles to the scene of an accident or where all private motoring is held up at Police road blocks.

Two methods are currently used:

(1) Magnetic or clamped roof racks saying 'water' or 'water emergency' or dayglo plastic windscreen stickers displaying a Wessex logo 'emergency vehicle'.
(2) The Authority provides from its operational workforce a small divisional team with a suitably designed vehicle that can be despatched to the scene of such an incident to assist in cleaning-up operations. This vehicle is equipped with booms of various sorts, absorbent material and neutralising chemicals if needed, and contains a wide variety of lighting,

tools and other equipment to deal with such situations. In appropriate cases breathing apparatus is carried in these vehicles and the staff trained in their use. Limited protective clothing is also available. The Fire Brigade have been particularly helpful in decontamination procedures and normally water authority staff involved in such an incident would be decontaminated by Fire Brigade personnel.

(3) Emergency equipment comprises:

Vehicles, oil booms, protective clothing, emergency lighting, maps/plans, breathing apparatus.

The water authority's laboratories are opened if necessary for the rapid computer based analysis of samples, tracing the source of a pollution, providing data for decisions to be made, for example when to shut river intakes etc.

14.6 COMMUNICATIONS

It is vitally important that when dealing with such an incident good communications are set up and maintained. This is sometimes achieved through the radio/telephone set provided in the car of the Water Quality Officer dealing with the incident, but if the incident is already reasonably large scale then a specialised emergency communications vehicle is despatched to the scene to stand alongside those of the primary emergency services. This has the advantage of being able to control onsite personnel working at or around the incident whilst communicating back to the divisional or Regional Control Centre. It also adds a focal point for the Police and Fire Brigade as far as the water industry is concerned at the scene.

14.7 EXERCISES AND TRAINING

It is vitally important that during the writing of Emergency Plans that realistic scenarios are prepared and both desk and limited field exercises are carried out to test all parts of the plan including manpower, plant and equipment and communications.

These exercises are not without their own problems; an exercise by the Wessex Water Authority some years ago involved the simulated crashing of a tanker containing domestic heating oil on a small bridge on Exmoor and staging the effects of discharging part of its contents into a small stream which was the head waters of a reservoir for public water supply in West Somerset.

This exercise was held in March and at the start of the exercise the weather was fine. In due course the appropriate code words were issued and the exercise started and Water Quality staff were immediately despatched to the site where one of the Authority's own supply tankers with suitable logos was positioned near the bridge where the spillage was taking place. Within half to one hour of the exercise starting however, conditions on Exmoor had changed considerably and the temperature fell rapidly and a blizzard set in. The high winds and driving snow caused chaos with communications and it soon became apparent that a number of the staff involved in the exercise were in difficulty. The Authority's communications vehicle was despatched to try to provide better communications and to discover what had happened. This vehicle had to turn back as it was unable to reach the scene in the snow and ice. The exercise was immediately called off and all efforts were put into rescuing the Authority's Water Quality staff who had been sent to the scene. This was effected after some time and one or two of the staff involved in the exercise were brought down from Exmoor suffering from the effects of cold. Although the exercise itself was a complete failure, some important lessons were learned both in terms of communications, personal equipment, vehicles etc., which we are sure would be of great value in similar circumstances!

Similarly, an exercise at the Authority's Black Rock Sewage Pumping Station at Weston-Super-Mare relating to a possible chlorine leakage resulted in this Authority changing its method of treatment as the exercise showed quite clearly that the timescales were such that no emergency plan could be envisaged that would effectively protect the population living around that particular installation, which in addition to residential areas included a school on the opposite side of the road and a proposed new general hospital.

After each incident a wash-up should take place so that lessons learned can be incorporated in emergency plans, the writing of which is a reiterative process. Similarly, the Wessex Water Authority carries out regular lecturing and training sessions with both the Police and the Fire Services on a wide variety of emergency situations. This has proved to have an extremely beneficial effect on the day-to-day working relationship between the Authority and the primary emergency services. Similarly, contact between divisional and regional staff and their opposite numbers in the Fire Brigades and Police forces is an ongoing situation and the value cannot be emphasised too strongly. This needs constantly to be referred and updated due to rapid turnover of staff, changes in telephone number etc., in all the forces concerned.

14.8 CONCLUSIONS

In addition to the emergency plans and operational practices of individual Water Authorities there have been a number of extremely valuable studies carried out in the last few years on oil and chemical spillages. One of the first of these was the oil and water industries working group which published an earlier report in 1972 which was widely distributed throughout the oil and water industries and local authorities in Great Britain.

The group in 1978 revised and updated their previous work and produced a booklet *Inland Oil Spills, Emergency Procedures and Action* in 1978 [1]. The membership of that committee involved the Institute of Petroleum, the Pipeline Industries Guild, the National Water Council, Scottish Rivers Purification Board Association, the Institution of Water Engineers and Scientists, central Government Departments and international organisations represented by the International Tanker Owners Pollution Federation. This booklet highlighted the statutory obligations in the necessity of good communications, the organisation and the emergency plans necessary for a successful clean-up operation and has specific sections on the characteristics of oils and their behaviour on land and water, and of the satisfactory containment or removal of such oils from watercourses. This report refers to an inland oil spill clean-up manual produced in 1974 by Stichting Concawe of the Hague [2]. This is the report of the Oil Companies International Study Group for Conservation of Clean Air and Water Europe. Between 1974 and 1979 they produced one or two reports dealing with oil spillage, clean-up manuals and the protection of groundwater oil and pollution.

These two documents proved to be extremely valuable when the National Water Council Directors of Operations Group was set up in 1977 under the chairmanship of R. J. Lillicrap, Director of Operations, Welsh Water Authority. The final report of this group, *Emergency Procedures on Pollution of Inland Waters and Estuaries* dated 1980 [3] has been widely used in the regional water authorities and is the basis for the operational approach currently used by the Wessex Water Authority.

As with most emergency plans the Wessex emergency plans for dealing with oil and chemical spillage are constantly under review and as new plans are refined and developed, they are exercised to test their effectiveness. There are still one or two areas of concern and constant liaison with Fire Brigades and other emergency services including the county councils and their scientific advisors is necessary.

The problem of protective clothing and in particular breathing sets for scientific staff of the water industry as well as the Police seems to be an area where some urgent action should be taken.

14.9 REFERENCES

[1] *Inland Oil Spills, Emergency Procedures and Action* 1978 — O.W.I.W.G.
[2] *Inland Oil Spill Clean Up Manual* 1974 — Stichting Concawe.
[3] Report on *Emergency Procedures on Pollution of Inland Water and Estuaries* 1980 — National Water Council, Directors of Operations Group.

CHAPTER 15

Contingency planning for chemical accidents

A. GILAD and J. I. WADDINGTON
World Health Organization Regional Office for Europe, Denmark

15.1 INTRODUCTION

Toxic and potentially toxic chemicals are now a part of the daily life of most people in the world. Because of the ever-increasing volume of chemicals that are being extracted, manufactured, transported, stored, used or disposed of as wastes, it is inevitable that accidents will happen with increased frequency all over the world. Considering the potential damage, both short-term and long-term, which the release of chemicals may cause to human health and environment, it is essential that every country should develop mechanisms for dealing with such accidents.

The ways in which different governments deal with accidents in general are highly variable and reflect the structure of government, culture and history of the different peoples in Europe. Most countries have some sort of emergency response system or set of systems for traditional accidents. However, the accidental release of toxic chemicals adds a requirement for information and expertise that most emergency response systems are not prepared for.

Any emergency response system is designed to reduce the impact of an accident by rapid containment. In the case of toxic chemical accidents, it is also necessary to know the nature of the chemicals, how to deal with them, the toxic, physical and chemical properties of the materials and the level of risk involved in contact, both for the emergency crews and the adjacent population. In this manner, the accident involving the release of potentially toxic or hazardous chemicals is different from a normal emergency (traffic accident, fire, train derailment, etc.).

No two accidents are exactly the same because, by definition, an accident is an unplanned event. Nevertheless, adequate contingency planning will help in prompt mobilisation of expert services, manpower and equipment to minimise the effects of accidents.

As the first step towards development of a response system for emergencies and accidents involving the release of toxic chemicals, the Regional Office had commissioned two background studies:

(a) organisational model of a countrywide emergency response system for chemical accidents; and
(b) survey of existing system components in European countries.

These two background documents were submitted to the Working Group on Contingency Planning for, and Response to, Emergencies and Accidents involving Potentially Toxic Chemicals, held in Bilthoven, Netherlands, 9–13 February 1981 and after review and revision, have been published in WHO/EURO Chemical Safety Interim Document series, to serve as a planning guideline

15.1.1 Objective

The objective of the guideline is to assist countries to develop compatible strategies for preparing contingency planning systems which are both adequate and rapidly and effectively implementable to respond to specific emergencies.

National adoption of this model will aid materially in developing a system which will be acceptable and compatible with existing emergency response systems and contingency planning programmes.

15.1.2 Types of emergency

Chemical emergencies may arise in a number of ways. Some of the more important types are as follows:

- disaster/explosion in a plant handling or producing potentially toxic substances;
- accidents in storage facilities handling large quantities of various chemicals;
- accidents during the transportation of chemicals;
- misuse of chemicals, resulting in contamination of foodstuffs, the environment, overdosing of agrochemicals, etc.;
- improper waste management, such as uncontrolled dumping of toxic chemicals, failure in waste management systems or accidents in wastewater treatment plants.

15.1.3 Level of accident

The method of classification chosen for this model is the scope or level of the accident. In this system a number of schematic contingency plans are provided which differ in scope and to some extent location of accident as well as in the ways of its handling. Four levels of accident considered are described as follows.

(a) *Level I (operator level)*

An accident where the adverse effects are limited to confines of one facility (such as a plant, railway station, storage depot, farm, gas or oil pipeline booster stations and/or terminals, etc.) and can be contained and controlled within that area by the operator on the site. It may require the resources of the whole facility to control, the effects of which are not expected to spill over into the community (Fig. 15.1).

(b) *Level II (local/community level)*

This involves an accident where the effects are spread to the public sector (the community) but can be contained by the resources of that community, plus resources of the plant or industry involved. The majority of transportation accidents, namely those occurring 'on route', will fall into this category (Fig. 15.2).

(c) *Level III (regional/national level)*

This may be a larger and/or more serious accident or it may be simply that it occurred at the border between two jurisdictions (regions or communities) within one nation or country. This may be described as an inter-jurisdictional emergency and may be handled with the resources available at the regional or national level, employing also the resources of the communities and industries involved (Fig. 15.3).

(d) *Level IV (international level)*

This is more than a complex accident exceeding the boundaries or resources of one nation. This may be a very large-scale national disaster or it may be a unique event requiring for its handling special skills or facilities not available in that country and/or it may simply be a small accident which occurs close to the border of a neighbouring country. The last type of emergency may be contained using national resources, but the management of the control may be undertaken by an international team (two or more affected nations) established for the purpose (Fig. 15.4).

The complexity of individual systems involved, as well as the variety of influences and conditions, require the preparation of contingency plans for each particular situation. It is possible, however, to utilise as a guideline a general model of contingency planning which respresents a systematic approach to the development of an emergency response system. Such a generalised system is shown in Fig. 15.5.

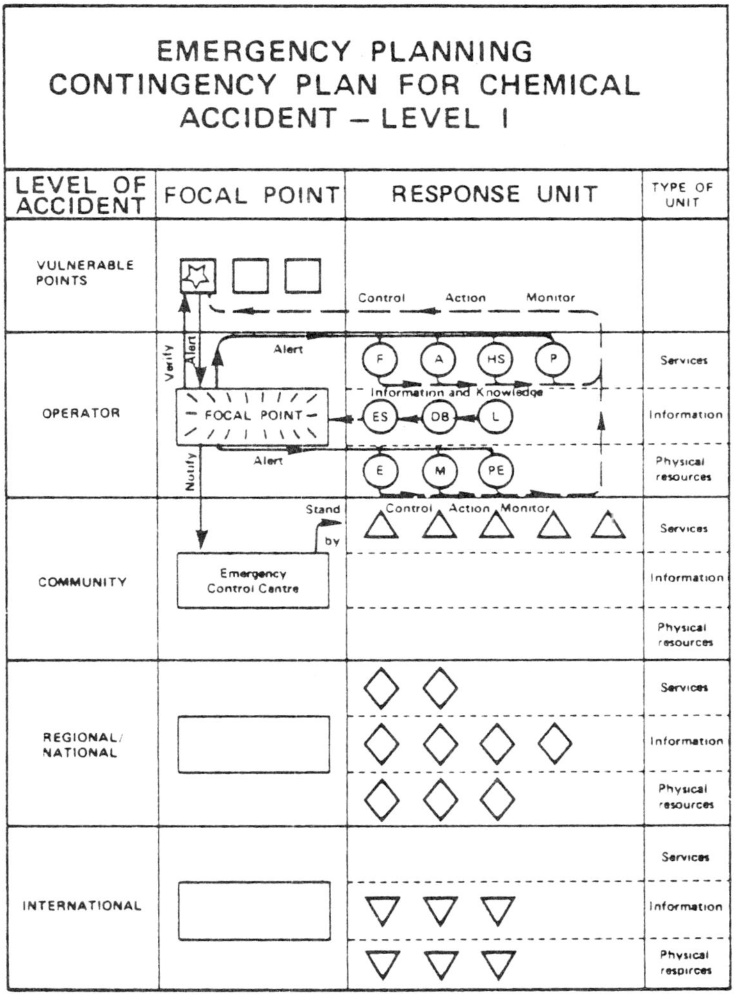

Fig. 15.1.

Legend

- ○ Internal resources of operator
- △ Resources of the local community
- ◇ Regional and/or national resources
- ▽ International resources
- AR Military services
- A Ambulances
- DB Data bank of information
- E Equipment
- ES Expert services (environmental)
- F Fire brigades
- HS Health services
- L Laboratories
- M Material
- P Police
- PE Personnel
- NS Environmental services, water waste management
- ——— Alert
- - - - - - - Information
- — — — Physical resources and action

Ch. 15]   Contingency Planning for Chemical Accidents   223

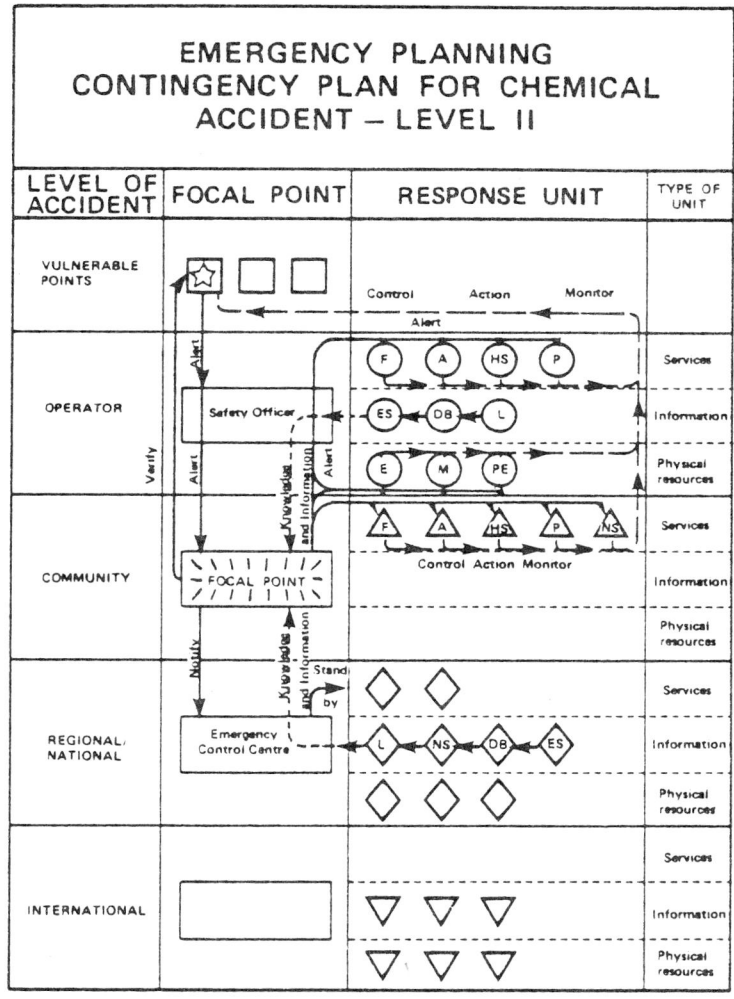

Fig. 15.2.

Legend
- ○ Internal resources of operator
- △ Resources of the local community
- ◇ Regional and/or national resources
- ▽ International resources
- AR Military services
- A Ambulances
- DB Data bank of information
- E Equipment
- ES Expert services (environmental)
- F Fire brigades
- HS Health services
- L Laboratories
- M Material
- P Police
- PE Personnel
- NS Environmental services, water waste management
- ——— Alert
- - - - - - - Information
- — — — Physical resources and action

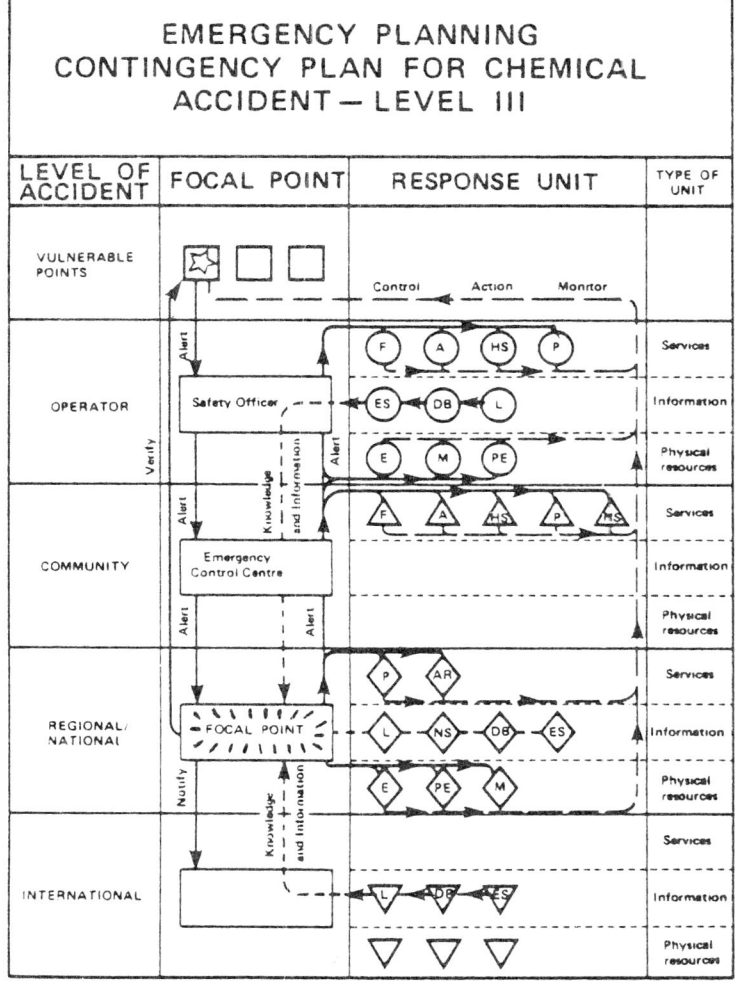

Fig. 15.3.

Legend
- ○ Internal resources of operator
- △ Resources of the local community
- ◇ Regional and/or national resources
- ▵ International resources
- AR Military services
- A Ambulances
- DB Data bank of information
- E Equipment
- ES Expert services (environmental)
- F Fire brigades
- HS Health services
- L Laboratories
- M Material
- P Police
- PE Personnel
- NS Environmental services, water waste management
- ———— Alert
- -------- Information
- — — — Physical resources and action

Contingency Planning for Chemical Accidents

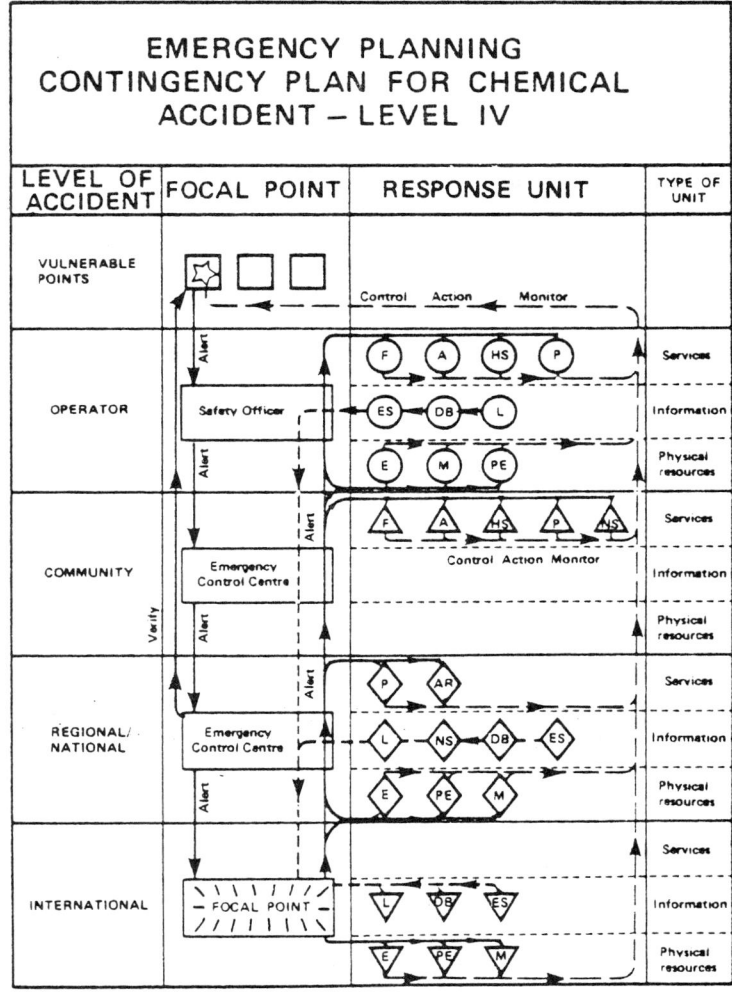

Fig. 15.4.

Legend
- ○ Internal resources of operator
- △ Resources of the local community
- ◇ Regional and/or national resources
- ▽ International resources
- AR Military services
- A Ambulances
- DB Data bank of information
- E Equipment
- ES Expert services (environmental)
- F Fire brigades
- HS Health services
- L Laboratories
- M Material
- P Police
- PE Personnel
- NS Environmental services, water waste management
- ——— Alert
- - - - - - - Information
- — — — Physical resources and action

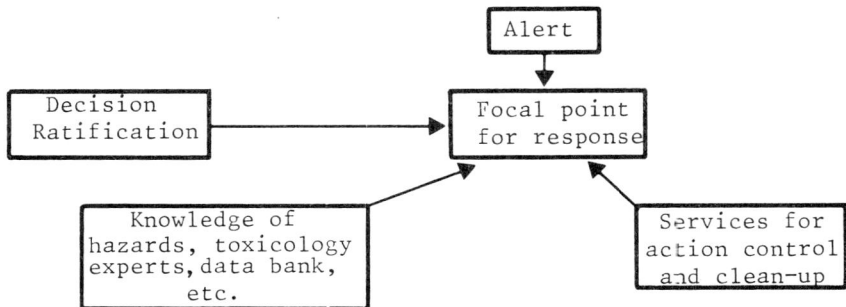

Fig. 15.5 — Generic management model for emergency response.

15.2 EMERGENCY RESPONSE SYSTEMS

15.2.1 Main objectives of emergency response systems

Accidents can occur in any industry and in almost any activity in spite of efforts to prevent them. In particular, those industries or activities dealing with toxic chemicals have a high potential for loss or damage. In those cases, the loss measured in both human and monetary terms may be severe. In some cases, because of effective and timely action, the loss has been considerably reduced. In these cases, effective action has been possible due to the existence of pre-planned and practised procedures for handling emergencies, utilising the combined resources of the operator and outside services.

The main objective of any emergency response system is to minimise any possible adverse impacts of accidents on people, environment and property. This requires the establishment of a system which makes optimum use of all available resources for speedy containment of the incident, protecting the health and safety of the prople, both nearby residents and workers, as well as minimising damage to the environment and property. The emergency response system must also provide adequate and accurate information to all relevant authorities and the public and provide ultimately for the safe rehabilitation of affected areas.

15.2.2 Systemic approach to the establishment of emergency response systems

If the contingency plan is to provide the expected results, that is significantly minimise the adverse impacts of the accidents, its development cannot be undertaken on an *ad hoc* basis, considering only selected parts of processes, operations or areas. Contingency planning should be a continuous activity, taking into account previous experience and both the present as well as future state of the industrial systems. This involves changing the emergency response system as changes occur in the processes, operations, products, plants and areas. The emergency response system must reflect the actual conditions, limitations and resources as they exist at any given time.

Contingency Planning for Chemical Accidents

The responsibility for contingency planning should rest upon an individual with assistance and advice from a committee established for that purpose. The final responsibility, however, should be clearly defined. The development of a comprehensive emergency response system should not imply that people responsible for planning this system should necessarily also operate the system.

The variety of operations, where an emergency could occur, is infinite. In each case, however, the system and its subsystems should be defined and possible synergistic effects investigated. This will provide a clearer picture of the probable impacts and consequences of potential accidents and provide a better basis for an emergency response system.

A comprehensive systems approach to the development of an emergency response system requires a process of contingency planning, including the following steps.

(a) *Investigation of vulnerable points, processes and/or activities*

At each level, a comprehensive analysis should be carried out on the most likely sources of accidents and emergencies. These may include some processes or installations in the plants, waste management facilities, transportation means and some parts of routes, some operations in agricultural farms, storage depots, etc. Detailed records of previous accidents or of near accidents would be of great help here. For each case, at least basic characteristic features of the area most likely to be affected should be investigated. These include population density, possible locations of the accident in relation to built-up areas, the prevailing wind, the possibility of contamination of air, water and soil, etc.

(b) *Estimation of possible chemical emissions*

The key information that is needed is the nature of the hazardous chemical or chemicals involved. An estimation should be made of what chemicals might be released, in what forms and quantities. Their classification may include flammable liquids and solids, combustible liquids, oxidising materials, explosives, etiological compounds, toxic chemicals, corrosive materials, etc. The emissions, as well as resulting ambient concentration, will depend not only on the processes and operations which may be involved, but also on the concrete course of the accident as well as on local climatic conditions. It would be useful to use, where suitable, existing simulation models, paticularly dispersion models, including a prediction of the changes in toxicity and concentration with time. The pattern of spreading the contaminants over the area would also be a useful tool to determine an appropriate countermeasure.

(c) *Knowledge of effects of toxic chemicals*

Recognising that pollutants may have a number of different pathways to cause health effects, it is clear that more information has to be available than a short list containing the chemical formula, the amount involved and the possible toxicity for man classified in a simple manner. For instance, due to influences of toxic substances on food chains or transport cycles through the environment, so-called minor emergencies may cause major damage in future.

It is obvious, therefore, that one of the most important parts of contingency planning and preparation for an effective emergency response is the detailed investigation of all the possible effects of toxic chemicals which may in any particular case be released into the environment. In doing this, one should pay attention not only to the original chemicals, but also to all possible products of their reactions during accident and/or during the process of dispersion in the environment. All this information should be properly classified, stored and maintained in a readily accessible form, together with instructions concerning where additional data may be obtained.

(d) *Knowledge of possible protective and remedial measures*

A preliminary investigation should be carried out on what possible protective or remedial measures might be taken to help minimise the adverse impacts of expected chemical emissions on the public and the environment. Local conditions, including the effect of season, weather, etc., should also be considered. The following lists, referring to an emergency due to chemical spill, may serve as an example of basic information required at this stage:

- the amount and properties of chemicals released into the environment;
- the direct danger to personnel in the area and possible protective measures;
- the nature of the damage and the possibility of repairs;
- the possibility of containment of the spill, for example, through a solid container, a form of dyke;
- the possibility of contact between the spilled chemical and waterways;
- the danger of explosion and/or fire;
- the effect of rain and wind on the spill;
- the equipment and materials necessary to restrict the distribution of the chemical on a larger scale;
- the possibility of spill treatment.

This analysis which is far from being complete might also provide a valuable feedback for the effective prevention of accidents.

(e) *Designation of responsibilities*

At each level, the responsibility for contingency planning as well as for implementation of an emergency response system should be defined, describing clearly the respective roles of the field commander (commanders) supervising the activities (at the site of the accident), and that of the focal point (in the emergency control centre).

Continuous services (24 hours/day) at crucial points and posts must be ensured.

(f) *Preparation of plan for action*

Based on the foregoing, the emergency response systems should be prepared on relevant levels as a standard operating procedure for handling of specific accidents as well as for fast response to any emergency. In preparation of the systems it is imperative to ensure consistency in contingency planning at various levels. Everyone involved in the implementation should be trained and the system should be periodically tested.

(g) *Establishment of the liaison with external authorities*

An important role in reducing the impacts of any accident is played by the emergency services of the fire, police, ambulance and health services, and occasionally by some units of the armed forces. The roles played by these agencies may vary from place to place. Liaison between services will help to establish what kind of assistance might be expected as well as ensure:

- effective co-ordination in control of operation;
- consistency of emergency response systems with any existing plans developed by outside authorities;
- that outside services are aware of the nature of the potential emergency;
- equipment provided at the site of the accident is compatible with that of outside services;
- personnel are acquainted with those with whom they are likely to be involved during the emergency.

(h) *Resources for handling the emergency*

Effective handling of emergencies requires the full utilisation of all available resources, internal as well as external. In order to facilitate this, sufficient funds must be available for this purpose. Furthermore, at each level continuous attention should be paid to adequate briefing and training of personnel, and the maintenance of equipment and material in a continual state of readiness is required. In some instances, particularly major emergencies, there is an acute need for the provision of external help of various kinds. Fast response to this need would be considerably facilitated by having available a set of up-to-date rosters, including:

- experts on specific problems: toxicity, technical aspects, remedial actions, etc.;
- institutions and laboratories which can quickly and reliably perform complex analyses and tests needed for decision making;
- organisations and firms which can provide trained personnel, equipment and material to effectively deal with the emergency.

(i) *Communication*

A system for collecting, processing, evaluating and disseminating information is vital for the proper operation of an effective emergency response system.

It must be ensured that adequate quantity and quality of information is received as well as transmitted to individual points of the pre-planned network, thus providing sufficient data to all parties involved, minimising duplication and gaps in information. Consistency of information to the public is essential in this regard. One person or agency should be responsible to ensure information released to the media and public is both accurate and internally consistent.

15.2.3 Structure and elements of emergency response systems

While contingency planning should be carried out and emergency response systems prepared for all installations, operations and activities having a risk of accidental release of potentially toxic chemicals, it is obvious that these systems will only be effective if they are properly integrated into a regional, national and, where appropriate, international structure. This need is conditioned not only by the fact that it is virtually impossible to predict the scope and consequences of an accident, but also the need to be able to undertake immediate protective measures in large affected areas and to facilitate the provision of all the necessary kinds of help.

It is necessary, therefore, to establish at each level of the country-wide structure a 'focal point'[†] to co-ordinate emergency response activities within the relevant area under their responsibility. The person named as the focal point shall also be responsible to process, transmit and receive all necessary information to the public administration and to the public and to facilitate the speedy provision of all necessary help.

While differing in particulars, depending on the scope of the emergency, the response systems should include the following elements.

(a) *The alert system*

This requires the establishment of suitable alarm systems, and there should be a standard procedure for passing the information on the occurrence of an accident to the appropriate focal point. The focal point may move up in the hierarchy depending on the level of the accident (see Fig. 13.5).

(b) *Evaluation of situation, classification of the accident*

Basic information must be provided enabling preliminary classification of the accident, probable consequences and actions required. This initial appraisal would normally be conducted by the person named as the focal point. At the same time, detailed data and information related to the accident are collected and internal, as well as external, expertise may be requested.

---

† The function of a focal point at operator level is often carried out by the 'safety officer' or 'chemical safety officer'.

(c) *Decision and alerting of the emergency response systems*

In most cases of accidents, time is a most important factor (explosions, sudden release of dangerous quantities of toxic chemicals into the environment) and decisions must often be based only on preliminary data and insufficient expertise. These decisions should be verified and corrected as soon as possible.

(d) *Provision of information*

Adequate flow of information must be assured to all relevant parties to ensure effective and fast response to the accident. This may be addressed to the following:

- operator's management, an industrial co-operative (for example Association of Chemical Manufacturers or government of relevant level);
- the focal point of a higher level;
- fire brigade;
- police;
- military or civil defence service;
- public health services, hospitals;
- construction firms;
- transport firms;
- public;
- others.

(e) *Provision of external help*

In many cases, accidents involving the release of toxic chemicals into the environment cannot be effectively contained and the adverse impacts minimised without some external help or advice. Some examples of this help or advice include:

- access to relevant information, particularly to toxicological data, allowing classification of the accident and the provision of qualified advice concerning protective equipment and remedial measures necessary;
- highly qualified assistance, which can provide from previous experience, and under the pressure prevailing at the time, information from dealing with the previous similar cases and an estimate of the probable consequences of the accident under the prevailing conditions;
- services of qualified, reliable and tested well equipped laboratories, to perform the necessary analyses and tests;
- provision of skilled personnel to deal effectively with the emergency;
- provision of such material and equipment as may be necessary to provide adequate protection and remedial measures.

In view of the serious consequences of a major emergency involving dangerous chemicals, a high degree of accuracy and reliability has to be assured in relation to the different classes of information that have to be supplied. The quality of data in relation to the detection and judgement on the seriousness of the emergency is especially important. Effective information systems should provide ready access to information tailored to the explicit needs of their users. In general, however, the existing data information systems do not contain all the information needed in emergency conditions. Extension and improvement of the information systems is urgently needed. To avoid one of the main problems, that is the fact that large computerised data bases always stay behind the current scientific knowledge, it seems to be useful to link together a network of small specialised data banks attached to scientific research institutes, where there are experts in the subject available to advise on the use, interpretation and maintenance of the data. This would also simplify continuous verification of the data validity.

Provision of reliable information in emergency situations requires not only highly qualified experts, but also selection of suitable laboratories whose reliability should be regularly checked. Apart from the ability to detect and measure the chemicals and to interpret the results, both experts as well as the personnel in the laboratories must be able to work under conditions of stress. This means that in the process of selection they should be tested also under simulated emergency conditions.

(f) *Decision on, and implementation of, required protective and remedial measures*

Even at an early stage following the accident, the provision of effective protective measures is of high priority. Simultaneously, possible future remedial or rehabilitative activities should be considered. This approach may prove itself in the future. The following actions may be required:

- evacuation of the plant;
- evacuation of the population from the affected area;
- organisation of receiving areas for the evacuated populations with adequate supplies and facilities;
- removal of the material spilled in the accident;
- in the event of a cut in water, gas or power supply, the provision of substitute safe and reliable supplies;
- changes of routes, for private and public transportation;
- provision of adequate food supply for emergency workers and evacuated people.

(g) *Continuous monitoring of the post-accident situation, adoption of relevant decisions and measures*

Monitoring and evaluating the public and environmental health impacts of the accident as well as the consequences of changes with time on the whole affected area is essential for effective handling of the emergency. The measures adopted previously must be modified as circumstances change in the specific situation.

(h) *Maintenance of communication links*

There are a number of parties to whom regular information on the development of the situation must be provided. These include, for instance, governmental authority at the appropriate level, health services, and the general public (preferably through a single information officer to the mass media to ensure internally consistent reporting).

(i) *Preparation of plans for rehabilitation*

Once the emergency is under control and principal causes and consequences clearly understood, a comprehensive plan should be prepared to ensure fast and effective restoration of both the environmental quality and the socio-economic activities in the affected area.

(j) *Post-accident analysis and evaluation of response activities*

When the emergency is over, it is desirable to carry out a detailed analysis of the causes of the accident, evaluate the influence of the various factors involved and propose methods to eliminate or minimise them for the future. At the same time, the adequacy of the contingency plan should be evaluated. Due attention should be paid to the efficiency of the emergency response system as well as the adequacy and timing of the various components of the plan.

(k) *Preparation of the final report on the accident*

The final report on the accident should provide a full picture of the accident, its causes, development, consequences, process of handling of the emergency, implementation of the emergency response system and the results obtained. The shortcomings of the contingency plan, failures experienced and successes achieved in preventive, protective and rehabilitative measures should also be recorded. The main objective of the final report is to record all experience and knowledge gained from the recent event to provide the basis for the future improvement of both the contingency plan as well as the development and practical application of emergency response systems.

15.3 PROTECTION AND REHABILITATION OF WATER RESOURCES

The WHO Regional Office for Europe, in collaboration with the Water Research Centre, is currently producing a guideline document on protection and rehabilitation of water resources contaminated by accidental spillage.

The guideline will define various strategies depending on the physical, chemical and biological properties of the contaminant and will deal mainly with technical aspects of clean-up operations of surface and ground-waters. Organisational, legal and economic aspects of the rehabilitation operation will also be discussed.

It is expected that the guideline document will be available before the end of 1983.

Discussion

Chairman: **A. G. SEMPLE**
Secretary, Water Authorities Association

The discussion was opened with a statement of agreement with Mr Harper's expression that 'prevention was better than cure'. Unfortunately, however, too many pollution incidents occur and water authorities do not have the resources to prevent all of them. In this context it was believed that a 'cure' effected by a prosecution of a polluter could have a large preventive effect. A friendly approach often proved to be less successful in preventing pollution than those instances where an authority applied its powers vigorously. The factors influencing the decision over whether to prosecute or not were discussed. These included the degree of pollution which has occurred, the attitude of a polluter and any previous history of pollution incidents. The problems involved were illustrated by citing the negligent attitude of one polluter, who was aware that he was causing pollution, but who was waiting for a 'verbal warning from the water authority' until he took any remedial measures.

Mr Harper had mentioned in his talk that under the Control of Pollution Act, the water authorities had the power to recover the cost of emergency action following a pollution incident. However, a delegate was under the impression that the part of the act giving water authorities this power, along with that which required flora and fauna to be restored following such pollution, had not yet been implemented. It was stated that, at present, some water authorities do try to recover the costs of these 'clean-up' operations, but that this was based on bluff rather than any statutory power.

Consequently, Mr Harper was asked if he thought that this was a signficant problem and also whether he thought that the Department of the Environment should accelerate the implementation of the legislation affording these powers?

Mr E. Harper answered by saying that his understanding was that a date had been set for implementation some time in 1985 or 1986. He thought that when such powers were given it could have considerably more financial impact upon polluters than the current penalties. North West Water Authority had been

Discussion

reasonably successful in retrieving these costs in the past by entering into a verbal or written contract with the polluter concerned who, if they were a large industrial company, generally had a responsible attitude. The presence of a statutory right to reclaim such costs would of course make the authorities' position much stronger.

With regard to Mr Harper's chapter, a speaker observed one fundamental difference of approach between the North West Water Authority and Severn-Trent Water Authority. This was that in certain rivers of low quality in urban areas, Severn-Trent believe that even if sewage and industrial discharges to these watercourses were improved, no improvement in riverine quality would be seen beyond Class 3, due to the impact of urban storm-water run-off.

Mr Harper responded by saying that he thought that there was no fundamental difference in the approach of the two authorities. With regard to the very small streams in urban areas with paved beds there were not going to be fisheries in such streams. However, the water quality in these streams may be good enough to meet the Class 2 criteria and this is an anomaly of the classification system. It was North West's intention that the larger rivers should support viable fisheries: this will be a great task in itself.

Clarification was requested on the relationship between the functions of river inspectors, trade effluent officers and fittings inspectors within the North West Water Authority referred to in Chapter 13. Referring to the Severn-Trent Water Authority it was stated that real benefit was gained in having an industrial water officer working in the River Pollution Control Section. In this role an officer could advise industry on the proper use of water as well as helping pollution control officers upon matters such as water balance.

Mr Harper answered by stating that the North West Water Authority was one of the few authorities left organised on the basis of Rivers Divisions. Thus the river inspectors and trade effluent officers had separate responsibilities but he emphasised that there was good liaison between them on various problems. There exists procedure within the authority's structure, which allowed all views on topics of general concern to be passed through one focal point.

Mr B. Alexander (North West Water) made the comment that water authorities should have a presence in sensitive areas in order to prevent pollution problems arising. Any of the aforementioned officers should be capable of identifying potential problems. In the North West Water Authority it was possible that the authority could take advantage of the fact that it had about 8000 staff covering the whole region, each of whom should be able to identify and cope with pollution problems to a greater or lesser extent.

Referring to pollution control it was pointed out that very often the parties most vocal in their concern over pollution are not the same people who are responsible for bearing the costs of environmental improvements. He thought that in their consultation procedure, the North West Water Authority were clearly trying to obtain the view of those interests most likely to bear the economic brunt of a proposed improvement programme. Information on how successful this approach had been in terms of soliciting responses from such parties was requested.

Mr Alexander acknowledged the difficulties on consultation due to the fact that at a baseline level every water-rate payer makes a contribution and therefore should be consulted. Clearly this was impractical, thus the authority attempted to make consultation as wide as possible, including discussions with elected bodies at different levels of government, even as far down as parish councils. Consultation with industry was also of great importance as alterations in consents could contribute considerably to costs in some cases. Consequently, the authority would attempt to give specific industries several years advance warning of any likely changes in consent and the impact of such changes is staggered if necessary.

The advent of consumer consultant committees in the near future was seen by Mr Alexander as another point of contact with the consumer as part of the authority's attempt to make consultation as wide as possible. A question was directed to Mr R. J. Whitaker (Wessex Water Authority) regarding his comment that his authority had water drainage maps for all the motorways passing through their area as part of their contingency plans for coping with accidental spillages. Mr Whitaker was asked if he knew if this type of emergency planning was in general use in the UK. Such a planning system had much to commend it and this point was illustrated by quoting an example of its use in Europe.

This concerned a survey which a company had carried out to assess the risks associated with the proposed path of a major oil pipeline which traversed an important aquifer for potable supply.

The aim of the survey was to evaluate the risks and potential impact of breakages or spillages from the pipeline at various points along its length upon the groundwater quality in the aquifer. The increased use of this kind of planning technique by the water industry in the UK in relation to pipelines, motorways and major industrial sites close to important groundwater resources was recommended.

Mr Whitaker recalled that the British Pipelines Association had used this type of planning in this country. He agreed that in terms of spillages on motorways it was very important to know where the spilled chemical was likely to go and his authority was currently discussing their contingency plans with the Ministry of Transport in order to examine the best method of reaching the site of any potential motorway spillage as quickly as possible. Another speaker was impressed by the Wessex Water Authority approach to spillages. He did however observe that such operations cost money. Thus he asked the authors of Chapter 14 if they had any figures for the costs involved in terms of the resources used.

Mr J. A. Young stated that although he had no figures available he would quite willingly produce them for inclusion in the conference proceedings. He also made the point that much of the resources in terms of vehicles and equipment were multi-functional and that the equipment and staff involved in the emergency operations reverted to another role when not on call. Some of the specialist equipment is only used in spillage emergencies, but this equipment is durable and lasts for a long time. Thus Mr Young concluded that overall the costs involved were not prohibitive. In a letter to the Editor, Mr Young provided further information:

> 'We estimate that Water Quality Officers who are usually involved in pollution incidents spend about 50% of their time dealing with such incidents, obviously on an ad-hoc basis. In many cases pollution incidents are costly to deal with and whenever possible we attempt to recover the cost from polluter's insurance. For example, the discharge of chlorine from a swimming pool into the River Avon involved re-stocking part of the river which took 6 hours of staff time. This cost of £1400 was paid by the polluter's insurers.
>
> In another case, an oil pollution occurred due to a ruptured feed line from a central heating tank. This incident cost £3800. After four months of negotiations this was paid by the oil company's insurers. However, in the case of a silage pollution where the silage had to be pumped from the river and the river oxygenated, this involved five days and nights of oxygenation at a total cost of £1744 but to date it has not been possible to get the polluter to pay for this work.'

The authors of Chapter 14 were then asked how much their approach was dependent upon obtaining information on the effects of chemicals; how it was obtained and was there enough of it? **Mr Whitaker** responded by saying that normally this information was obtained by consultation with various members of staff within the authority. In general this information proves to be sufficient, although the simplistic answer is that one can never have enough information.

Mr A. Gilad stated that he doubted that water authorities would ever have enough experience, or knowledge of the toxic effects of many of the chemicals likely to be involved in such spillages, unless they had an extensive scientific back-up. This information was vital in choosing the best option for dealing with the accident. Continuing the topic of spillage control the conference's attention was drawn to the fact that, in the North-East, ICI had emergency vehicles which were deployed when accidental spillages occurred to supplement the emergency services in the North of England.

Another delegate from Wessex Water Authority agreed that obtaining information quickly on spilt chemicals was a difficult process. However, the most valuable source of this information in his opinion was the Water Research Centre, where a computerised database is maintained. There was a suggestion of running down this service and the hope was expressed that this would not happen because of the very useful role it fulfils.

Mr Gilad was asked if it would be possible for the WHO to compile an annual register of all major chemical accidents, as this would make people aware of potential problems and thus enable preventive action to be taken and contingency plans to be drawn up. An example was given of a fire involving PVC liquid that occurred on a Friday evening. Later on it was found out that a similar incident had been dealt with in Germany some time earlier and knowledge of this would have been most beneficial.

Mr Gilad agreed that such a record would be very useful and indicated that his organisation would definitely attempt to collate the information. Mr Gilad did note, however, that the WHO definition of Europe encompassed several countries from whom it might be difficult to obtain the relevant information.

CHAPTER 16

Detection of pollution at drinking water intakes

G. P. EVANS and D. JOHNSON
Water Research Centre

16.1 INTRODUCTION

Lowland river intakes currently provide 30% of the water supplied in the UK. On some estimates this will nearly double by the end of the century to about 7500 Ml/d while the number of river intakes will increase from the present 132 to about 200. This source of drinking water is very much more exposed to the possibility of pollution than the two other main sources: upland reservoirs and groundwater supplies. Figure 16.1 illustrates the range of possible sources of point-pollution. Identification of such sources in the catchment upstream of the intake will indicate the nature of most likely pollutants and thereby establish a basic set of parameters which should be monitored. At about half of the intakes intermediate storage is provided prior to treatment; this acts as a buffer against acute changes in water quality. However, for the rest of the intakes no such storage exists and water may be in the distribution network on its way to the consumer only 4 hours after abstraction from the river. About 6000 pollution incidents are notified in England and Wales each year and annually there are about 100 intake closures due to pollution incidents in the supplying rivers. The pollutants involved in the closures may be categorised as in Table 16.1.

Table 16.1
Pollutants involved in the closures of intakes

Pollutant	Percentage occurrence
Fuel oils	40%
Farm waste	10%
Sewage	10%
Plating effluent	5%
Phenols	5%
Other chemicals from industry or road accident	30%

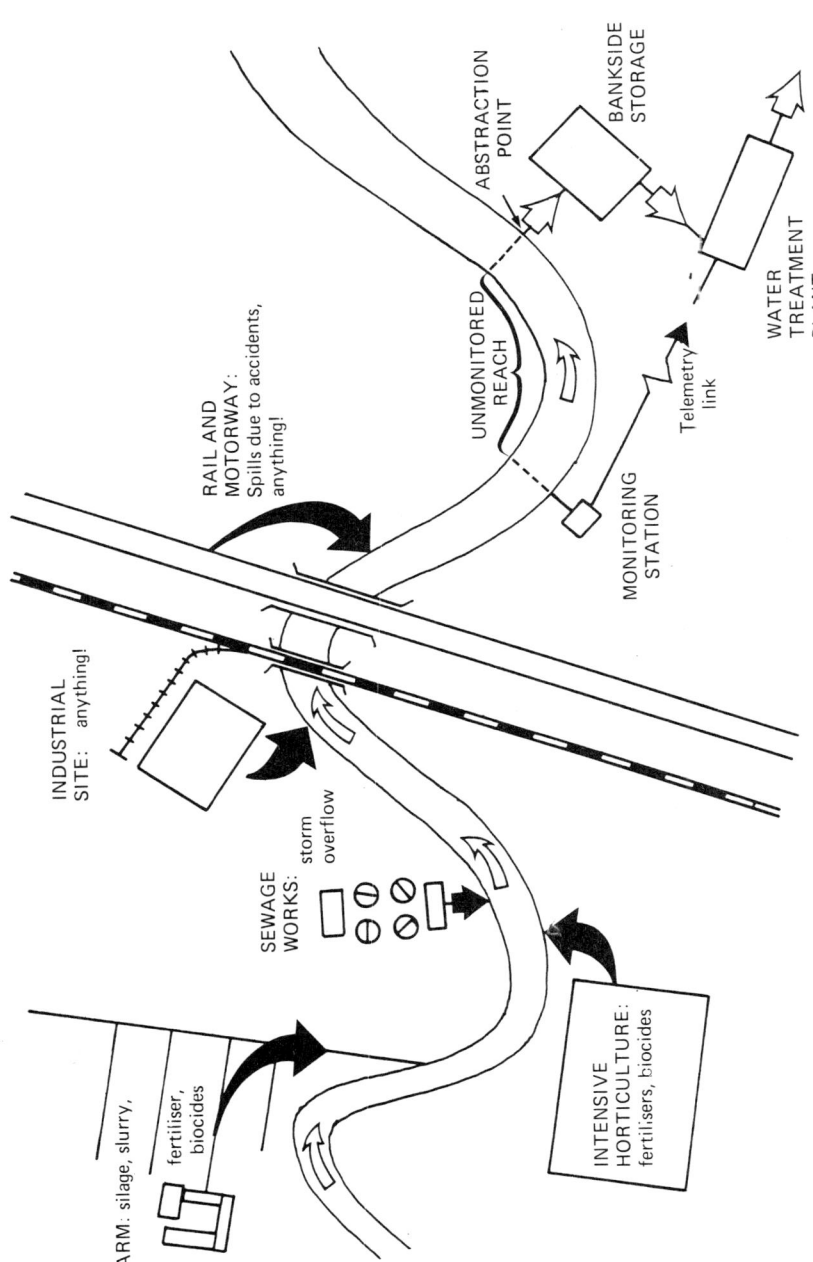

Fig. 16.1 — Sources of point pollution, water treatment plant and monitoring station.

Ch. 16]   **Detection of Pollution at Drinking Water Intakes**   241

In addition to these there are typically 20 closures a year as a result of seasonally high nitrate concentrations; this is primarily a problem in highly intensive agricultural areas, for example East Anglia.

As can be seen from the table the pollutants cover a very wide range of chemical types, some of which can be directly detected and others by inference. However, the list would include many substances for which there is no easy method of detection unless precise information regarding the incident has been given to the water authority. It is highly likely that in addition to the notified events there are many minor and maybe some major pollution incidents, accidental or deliberate, of similar importance to those listed above. There is no evidence that positive harm to the consumer has resulted from these incidents but those which are not reported until detected at a consumer's tap and which require subsequent interruption of supply do nothing to improve the confidence of the customer.

Certain pollutants, though not dangerous to the consumer immediately, can impair the treatment process so that the consumer eventually receives inferior quality water. The impairment of the process may last some time; for example, the biological layer of a slow sand filter may take several days to recover from a gross pollution incident. Even a modern physico-chemical treatment plant will suffer if exposed to oils and organics. Control of residual levels of chlorine may be affected by the presence of ammonia in the raw water. While some of the pollutants involved are potentially toxic to the consumer, those that are not can result in unpleasant taste and odour problems in the distribution system especially following chlorination.

The early detection of pollution incidents will allow the manager of the water treatment plant to make appropriate decisions in order to safeguard both the consumer and the plant. A good rapport with police and fire services will encourage the prompt notification of known incidents. For comprehensive protection, a continuous watch must be kept on the quality of water in the river so as to include unreported incidents. Ideally the monitoring position would be at some short distance upstream from the abstraction point so as to provide some degree of forewarning. The optimum position would be determined by consideration of the flow velocity of the river, diffusion and dispersion rates in the river, the transit time of the water through the plant and the location of any likely point-pollution sources. Details of suitable methods for calculating or measuring time-of-travel and dispersion properties of the river can be found in White and others [1].

16.2 CONVENTIONAL POLLUTION AND INSTRUMENTATION

Continuous monitoring of such parameters as dissolved oxygen, temperature conductivity, pH, ammonia and nitrate, organic matter, and turbidity is often used for optimum operation of the water treatment process by feed-forward control. Gross variation in these parameters can be taken to indicate the possibility of a pollution incident having occurred in the river. Such instrument packages are readily available from several manufacturers at moderate cost. Several parameters may be measured with specific ion sensors; these include ammonia, cyanide and nitrate. Although nominally specific the response of these sensors can be subject to either positive or negative interference from other ions. Furthermore some of these devices must be operated under carefully controlled conditions. For example it is often necessary to accurately control pH to values outside the range normally encountered in natural waters and sometimes the water to be sampled must be maintained at a known fixed temperature. Ancillary equipment is needed to maintain the required operating conditions which may be different for each electrode. Complete systems for ammonia determination are now available 'off-the-shelf' from several UK manufacturers; such systems will include a sample pump, pre-treatment chemical pump(s), constant temperature water bath and the necessary electronic instrumentation, for example for control of automatic calibration against high and low standards.

The signals from such instruments may be digitised and supplied to a microcomputer for automatic control and warning purposes. Nevertheless a commitment to instrumentation requires a commitment to maintenance. All of the probes are liable to biological fouling; this will be immaterial for some measurements such as that of temperature so long as the through-flow of water is not impeded but sensors involving the use of semi-permeable surfaces such as dissolved oxygen and pH sensors will be affected both by an increase in response time and also localised biological processes. It will be necessary therefore with existing instrument packages to instigate a soundly based programme of regular maintenance which will probably include re-calibration. Unless there is duplication of the sensor heads and associated plumbing, the maintenance process will render particular parameters temporarily unavailable.

The required performance of instrumentation for intake protection is rather different from that of instruments used in the laboratory for the accurate measurement of concentrations of particular determinands. For intake protection it is only necessary to indicate the presence of a determinand at a concentration deemed to be harmful either to the treatment process or to the consumer, whereas the analyst in the laboratory expects a response from the instrument which is a continuous function of the concentration of the determinand. As a minimum specification an intake protection system would monitor a specified range of prescribed substances for exceedance of defined thresholds; if this should occur for any of the substances an alarm would be generated with an indication of the pollutant responsible.

16.3 TWENTIETH-CENTURY AGRICULTURE AND INDUSTRY

The intensification of agricultural practices in both livestock and arable farming has resulted in an increased risk of point pollution from what were previously considered to be diffuse sources.

The import of fertilisers, their bulk storage and concentrated application increases the likelihood of inorganic pollution from nitrate and ammonia. The storage and use of herbicides and pesticides poses an especial problem since serious pollution can be caused by low concentrations of these substances.

In livestock farming, changes in overwintering methods necessitate the storage of large quantities of silage for winter feed and the resulting waste products in the form of slurry. Stored silage produces a highly concentrated nutrient liquor, slurry is rich in ammonia; leakage of either of these liquids to a water course constitutes a serious pollution event. Slurry spread on fields in inappropriate weather can result in very rapid runoff of nutrient.

Combinations of the instruments described in the previous section can be used for the detection of silage and slurry leakages. Such events are as serious as pollution from untreated sewage.

It is well known that many of the biocides used are highly toxic to human life but they will not cause a response in existing physico-chemical monitoring systems and there are no specific ion electrodes for their direct measurement.

New chemicals for use in both agriculture and industry are being developed every year and, although analytical techniques are available for their determination, these tend to be time-consuming, require sophisticated analytical equipment and so are wholly unsuitable for on-line application to intake protection. Therefore, to afford maximum security, the intake protection system should not merely be capable of detecting dangerous concentrations of anticipated pollutants but also able to indicate the presence of other unforeseen contaminants. Even the list of anticipated pollutants may be very long so that separate determination of their concentrations would require extensive and thus expensive instrumentation; it becomes clear that a successful intake protection system needs an instrument which will respond with a single alarm to many different types of pollution. Such an instrument is generally known as a broad-band monitor.

16.4 THE USE OF CAPTIVE FISH AS A BROAD-BAND MONITOR

One possible way of detecting the broad spectrum of toxic pollutants is to use a biological system along the lines of the coalminer's 'canary in the cage'; the reasoning behind this approach is that the response of the chosen biological system is expected to be analogous to that of the human consumer. For protection of water supplies such a system should be at least as sensitive as humans, reliable, easily maintained and show a rapid response which is amenable to automatic instrumentation. A monitor based on aquatic organisms is the obvious choice since these are well known to be sensitive to changes in water quality. Fish have been shown to respond to stress in the form of pollution by exhibiting one or more of: increased activity, avoidance reaction and changes in heart and ventilation frequency. These responses have been exploited in a number of monitoring systems of various degrees of automation using a wide range of fish species. Examples of such systems are the Poels avoidance monitor [2,3]; the Kerren Aquatest Fish Monitor[†] which detects change in the rheotactic response of fish in the presence of a pollutant; the Zippe-Bio-Control-System[‡] which uses the naturally emitted low frequency electrical pulses of Nile Pike (*Gnathonemus petersi*); the Petry monitor which uses magnets implanted in fish to indicate the degree of fish movement; and the work of DeGraeve [4] and Morgan and others [5].

One response which is receiving increased attention is the change in ventilation frequency. This technique makes use of the fact that the musculature involved in ventilation produces a small oscillating voltage in the water. The results of experiments carried out using a number of different pollutants and with several different species of fish are reported in the literature [6–14]. Methods of signal detection and processing by computer systems are being developed [15, 16] and systems are in use for the monitoring of industrial effluents [17, 18]. An appropriate species for use in the UK and Northern European environment is the rainbow trout (*Salmo gairdneri* Richardson) which is inexpensive and readily available from fish farms. The ventilatory signal can be detected by electrodes immersed in the water but not attached to the fish; for trout 15 cm in length, the amplitude is of the order of 40 to $100\mu V$.

16.5 THE WRC MK III FISH MONITOR

The monitor used at the Water Research Centre (WRC) for application to an on-line Intake Protection System is a refinement of an earlier model designed at the Centre [19] and is able to monitor separately up to 6 fish. Each fish is contained in a separate tank equipped with its own set of disc electrodes made from a sintered compound of silver and silver chloride. A description of the electrodes may be found in Solman [20].

The amplified signal from a fish may be traced on a chart recorder; ventilation frequency may be deduced by counting cycles and any peculiar behaviour of the fish, such as 'coughing', may be noted. A rise in ventilation of frequency or an increase in the frequency of coughing may be used as an indication of fish distress. But clearly such a method requires human intervention and, while it is feasible in the laboratory for the assessment of the effect of pollutants in controlled experiments, it is unsuitable for running automatically at a remote site.

[†] Kerren Kunstoff Technic. W. Germany.
[‡] Zippe GmbH + Co. Postfach 370. D-6980 Wertheim.

Analogue electronic methods could be used for counting cycles but the warning system as a whole requires a degree of automation not easily reached with analogue equipment. A microcomputer is essential for making decisions which may lead to warnings and it is convenient to use that machine for the original analysis of the data. This requires analogue to digital conversion of the data after suitable amplification.

Ventilation frequencies may be estimated by cycle-counting on the stream of digitised data but the method is prone to interference from other components of the signal, for example mains hum which cannot always be reduced to satisfactorily low levels. A more robust technique is the Fourier Transform method; this method requires the storage of considerable quantities of data for short periods but this is now easily possible with the greatly increased power of modern microcomputing systems. The ventilation frequency is shown as a peak in the power spectrum derived from the transformed series; this peak is easily distinguishable from the peak due to mains hum.

A range of credible fish ventilation frequencies is defined, typically 0.3 to 3 Hz. The spectrum is scanned to determine whether within that range there exists a peak which is greater than some predefined threshold. If there is no such peak, no estimate of ventilation frequency is deemed possible, (for convenience it is set to zero).

The response of the Mk III Fish Monitor to various pollutants has been evaluated in the laboratory. Figure 16.2 is an example of a pollutant test. The test lasted 4 hours and the fish were dosed with γ-HCH at $60 \mu g/l$ over a 1-hour period which started about 90 minutes from the beginning of the test.

The upper part of the figure shows plots of the ventilation frequencies estimated at 1 minute intervals for all of the fish except fish 4 for which signal-processing was temporarily unavailable. It can be seen that the pre-dose frequencies vary between fish but in general remain constant. Reaction to the dose starts after 10 to 15 minutes of the start of dosing and takes the form of either increased variability or a trend or both of these. The figure clearly shows to the unaided eye that the presence of the pollutant has affected the ventilation frequencies of the fish. Similar responses have been detected with ammonia, phenol, and paraquat.

For use in an automatic warning system a well-defined algorithm for detection of significant changes in ventilation frequency is required. One simple method, which appears to work satisfactorily when applied to data derived from laboratory experiments, compares a set of recently obtained ventilation frequencies of the fish (the inspection window) with a set of frequencies obtained an hour previously (the background window). The number of recently obtained frequencies lying outside the range defined by the 5% and 95% limits of the older data is used as a measure of change in behaviour of the fish. The signal processing and the statistical technique are described in more detail in Evans and Johnson [21].

The result of this analysis of the data is shown in Fig. 16.2 as a series of

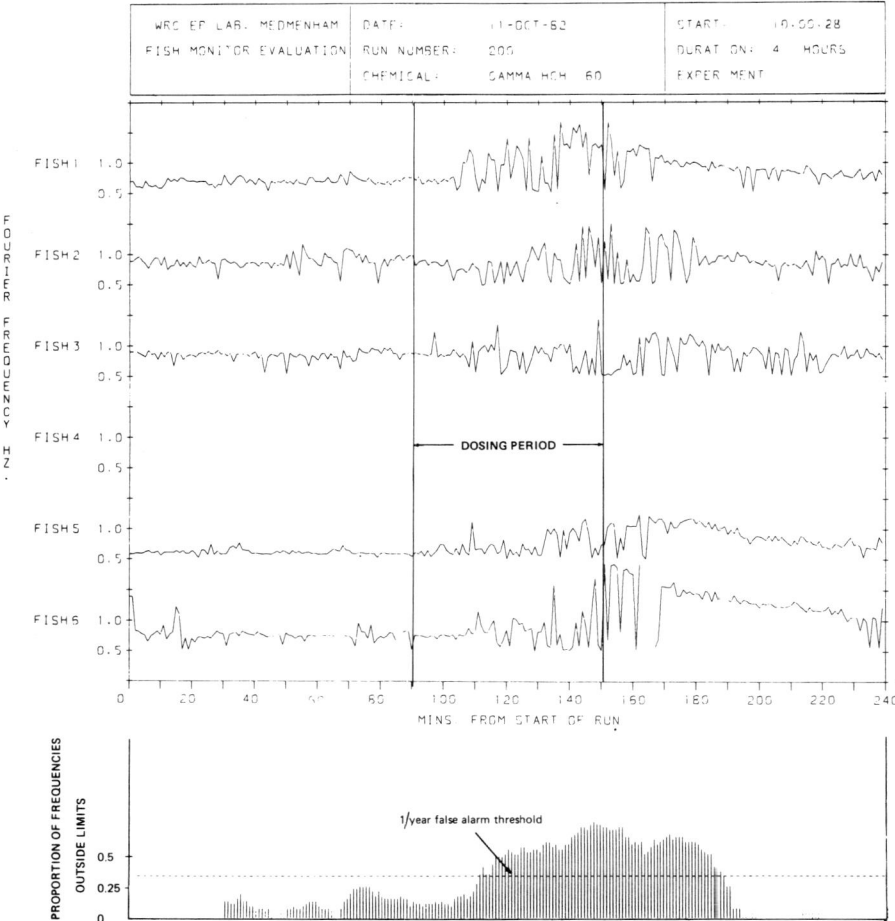

Fig. 16.2 — 4-hour pollutant test run: 60 µg/l γ-HCH dosed for 1 hour.
Upper graphs: Ventilation frequencies of five fish.
Lower graph: Alarm detection — proportion of frequencies outside limits, 1/year false alarm threshold indicated.

bars in the diagram below the frequency plot. The height of each bar indicates the ratio of the number of ventilation frequencies in the inspection window lying outside the 5 and 95 percentile limits to the maximum possible.

Also marked on the lower diagram is a dashed horizontal line corresponding to the ratio which one might expect to be exceeded once a year by chance. It can be seen that this ratio was exceeded over a prolonged period during this test.

Ch. 16] Detection of Pollution at Drinking Water Intakes 247

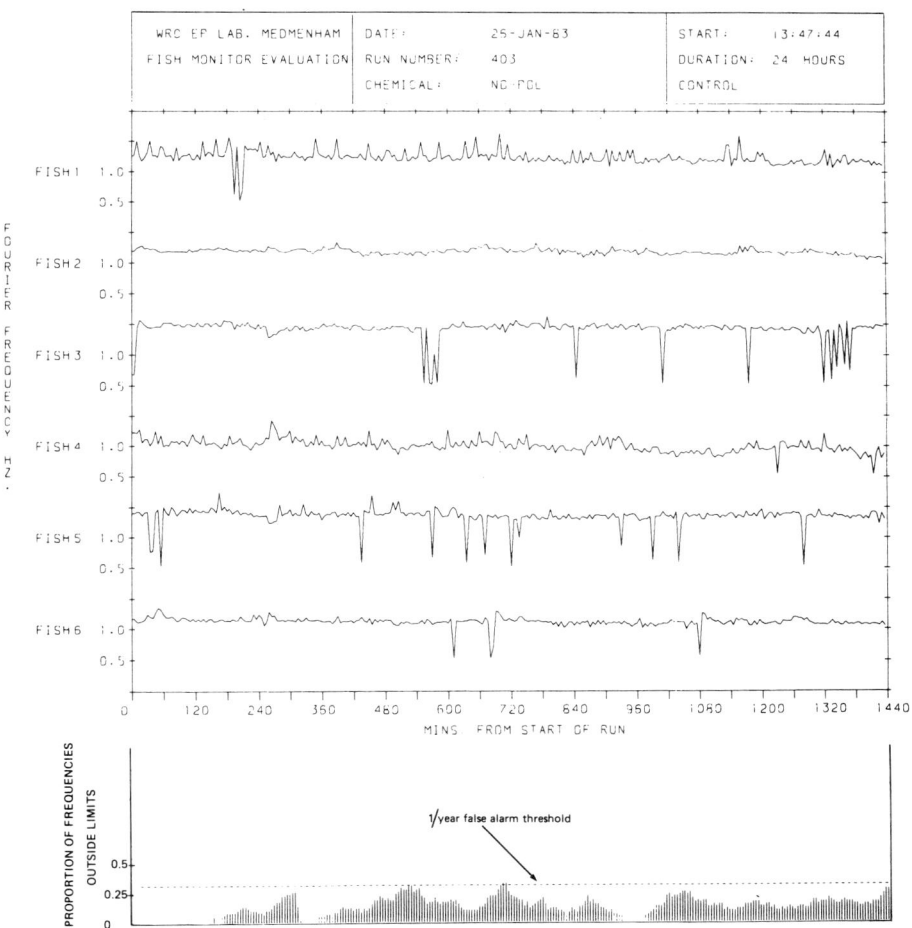

Fig. 16.3 — 24-hour control run.
Upper graphs: Ventilation frequencies of six fish.
Lower graph: Alarm detection — proportion of frequencies outside limits, 1/year false alarm threshold indicated.

In contrast, Fig. 16.3 shows plots of the ventilation frequencies of 6 fish estimated at 5-minute intervals over a 24-hour control period, during which there was no dosing; the lower diagram shows the ratio of excursions to the maximum possible. The ratio just exceeds the threshold on two occasions which, because they are isolated, can be handled by a suitable decision algorithm.

Table 16.2 lists the minimum pollutant concentrations which the Mk III fish monitor has been capable of detecting. The concentrations which produce responses are considerably greater than the EEC guideline concentrations shown in the table. However, the concentration-time profiles of spillages (which will constitute the major sources of acute pollution) rise rapidly; so the time required to reach the trigger concentration will be short and the alarm can be raised before the pollution front passes through the treatment works.

Table 16.2
Response of the Mk III fish monitor to pollutants

Pollutant	Concentration producing significant response	EEC maximum admissible concentration (MAC)
Phenol	3.75 mg/l	0.5 μg/l
Paraquat	1.6 mg/l	0.1 μg/l
γ-HCH	2.6 μg/l	0.1 μg/l

It is the intention that further tests will be carried out to determine the detection limits of the Mk III monitor to new priority pollutants as they are identified by the water industry.

16.6 DEVELOPMENTS OF BIOLOGICAL METHODS: WHOLE CELLS, ENZYMES

The use of whole organisms requires pieces of equipment which are large in comparison with currently available physico-chemical monitors. In addition individual organisms of the same species demonstrate different levels of response when exposed to the same concentrations of pollutant. The latter problem can be minimised by using several organisms from the same species and combining the information from their responses as described in the preceding section; this of course is at the expense of increased size and complexity of the apparatus.

The difficulty associated with the variation of response of different individuals can be reduced by descending the ladder of biological complexity from organism to single cell and even from cell to the sub-cellular level. At this latter point the response of the biological system becomes a defined chemical process. This approach offers the possibility of not only greater reproducibility of results but also a considerable decrease in size of equipment. This has had important implications in the biochemical and medical fields where most of the advances have been made. Acknowledged world leaders in this area are the Japanese, see Suzuki *et al.* [22].

Just as it is necessary to retain a large organism within flow-through monitoring equipment, for example a fish must be confined to the swimming chamber in order that its response may be observed, so must the components of a microbiological system be kept within the apparatus. Because the components of such a system are so much smaller, special methods have to be used; these include gel-encapsulation or surface adsorption of enzymes, or the retention of whole cells between semi-permeable membranes. Keeping the biological material within the instrument is a branch of biotechnology known as 'Immobilisation'. In the field of food and drug production immobilisation allows the re-use of expensive enzymes; in a similar way in bio-sensors immobilisation prevents the flushing away of the cells or enzyme and also in some cases provide a means of directly extracting a signal from the biological process.

Some organic pollutants may act as a substrate for certain enzymes or whole cells so that the arrival of a suitable pollutant will increase the activity of the system; in other words the method of pollution detection consists of looking for a positive response from the system either in the form of an increase in product or some electro-chemical phenomenon. The positive response or enhancement method requires the holding of the cells or enzymes at very low levels of substrate in order that, when the pollutant does arrive, maximum response occurs to this new supply of substrate. Deterioration of cell systems may occur during the 'starvation' period when the water does not contain the relevant substrate. Whether whole cells or enzymes are used, natural background levels of substrate may maintain the biological processes at such a level that pollution does not produce a significant increase, in other words the systems may not be sufficiently discriminating.

More commonly pollutants act as inhibitors rather than as substrates to enzyme-mediated processes and so an alternative approach is to choose cells or enzymes whose activity is depressed by the presence of one or more pollutants; pollution is thus indicated by a reduction in activity. In this case the system must be continuously fed with an appropriate substrate added to the incoming water so that the norm, uncontaminated water, is represented by a high level of response. The repsonse is quantified either electro-chemically or in terms of the amount of product.

This technique has been exploited by Holland and Green [23] in a gross pollution detector based on the inhibition by pollutants of the nitrification in biological columns. A natural population of cells was allowed to grow on granite chips sprinkled with river water. When the population had become stable they began to dose the column with ammonium chloride. Ammonia concentrations were measured with a specific ion electrode before and after the column, a large difference indicating success of the nitrifying process, a small difference some degree of inhibition. Stroud and Jones [24] used such a monitor with some success in field trials on a number of rivers in the UK Midlands. A version of the gross pollution detector is incorporated in Severn-Trent Water Authority's Holt Fleet automatic monitoring station [25].

Other examples of the use of whole cells for detection of pollution by inhibition are the Beckman Microtox system, which exploits the suppression of light emission by luminescent bacteria when exposed to certain pollutants, and also the Danish work at the National Agency for Environmental Protection, which employs the suppression of uptake of oxygen by a species of *Geotrichum*.

The inhibition by pollution of enzyme activity has been employed by the Midwest Research Institute portable pesticide monitor known as CAM-4 which has been shown to detect sub-toxic levels of nine commercial pesticides.

The Water Research Centre has commissioned research by UK organisations into whole cell and enzyme methods for pollution detection. Avenues explored have included inhibition of alpha-amylase and protease enzymes by biocides and the use of the methanotroph *Methylosinus trichosporium* for detection of organic compounds by positive response. There will be further development towards direct electro-chemical detection rather than measurement of product concentration or substrate consumption.

16.7 OTHER SYSTEMS UNDER DEVELOPMENT

Oil and hydrocarbon spills all sources account for 70% of the reported pollution incidents in UK rivers, so an instrument designed specifically for detecting such pollution would be of considerable use. The most obvious indicator of the presence of oil is a reflective layer on the surface of the water. An instrument is marketed by Camlab of Cambridge for the detection of such layers with the use of a back-reflected laser beam. Such a technique will only detect an insoluble fraction at the surface; since water is often abstracted via a submerged intake this floating film can be prevented from entering the works. However, of more consequence is the water-soluble fraction of any oils which may be present and it is much more desirable to detect these constituents. To this end the Centre commissioned the development of an oil sensor by the MidWest Research Institute, Kansas, USA. The instrument, the Strippable Pollutant Monitor (SPM), now undergoing proving trials at the Centre, works on the following principle. Clean air is bubbled through the water under test where it takes up volatile fractions of oils from the sample. The air then passes over a set of Stannic Oxide semi-conductor sensors whose conductivity increases with the adsorption of oil vapours. Initial tests showed that the device responded significantly to petrol concentrations as low as $0.1 \mu l/l$ ($60 \mu g/l$). Tests will be carried out to determine whether its performance with less volatile oils such as lubricating oil can be enhanced by pre-heating the flow-through sample.

16.8 SYSTEM INTEGRATION AND HOW TO USE A WARNING

No single instrument can provide comprehensive warning of pollution at an intake and so a complete intake protection system requires the concurrent operation of several instruments. In the simplest case the readings of the various instruments can be relayed, displayed and maybe recorded on chart at a central control room. Individual displays or chart recorders can be fitted with alarms and the choice of

Ch. 16]   **Detection of Pollution at Drinking Water Intakes**   251

appropriate control action can be left to the control room supervisor who will interpret a warning from one instrument in the context of the readings of other parameters.

A microcomputer under the control of real-time software may be used to organise the collection of digitised information from various instruments. The computer will display values of the measurements either as hard-copy or on a visual display unit. Warnings may be shown in the event of any specified limits being passed and may be supplemented by audio-alarm. Examples of such integrated systems include Sunderland and South Shields Water Company's station on the River Wear [26], and Severn-Trent Water Authority's system at Holt Fleet described by Freeman [25].

For those treatment plants where there is no 24-hour staffing the amalgam of information from different instruments cannot be directly accessed by a human on site; however alarms can be automatically delivered and values telemetered to a central control room or to a duty officer. Modern microcomputer technology extends the opportunity for an automated intelligent assessment of the information at the measurement station prior to alarm; this may improve the discrimination between true and false alarms. The stage has not yet been reached where a decision for appropriate control action can be automatically taken but it is possible for a properly programmed microcomputer to 'decide' whether to autodial an alarm call to a duty officer.

Regardless of how any automatic warning is communicated to the responsible staff, the right control action will depend on individual circumstances of the river and treatment plant. Communication with the police and fire services may establish the location of a spill while reports may have been received from anglers and others who keep an unpaid watch on the river. This information in conjunction with the known flow velocity of the river at the time will serve to establish the likely duration of grave pollution at the intake. Bankside storage allows time for more detailed chemical analysis to be carried out and additional information gathered. The decision as to whether to stop abstracting and to draw from reserves in the bankside storage or whether to rely on dilution in the storage basin will depend on the seriousness and expected duration of the pollution event in the river as well as on the retention time of the bankside storage.

It is obvious that without bankside storage between intake and plant any pollution incident has more immediate and serious implications; in the worst case the only recourse may be to terminate supply to the consumer.

16.9 WHERE TO NEXT?

The Water Research Centre has designed a Mobile Intake Protection System (MIPS) for the demonstration and evaluation of conventional and novel instruments in the field. The MIPS will be made available for use at a range of water intakes and treatment plants for which the contributing catchments are prone to pollution of varying degree and type. The system includes a microcomputer which collects and assesses the information from the instruments and is programmed to issue warnings in the event of pollution. An important part of the development and evaluation of the system will be the establishment of reliable software adaptable to the needs of individual intake sites.

The MIPS will incorporate the latest version of the fish monitor. The Water Research Centre intends to issue to the water authorities a design specification for the fish monitor as the basis for tender by commercial manufacturers. Testing of the effects of pollutants on fish will continue at the Centre and the results will be communicated to the water authorities.

The Strippable Pollutant Monitor is to be developed as an on-line instrument and will be eventually incorporated in the MIPS; a specification will also be made available for this instrument. The Centre will continue to follow closely progress in the field of whole cell and enzyme sensors and encourage their application to the detection of aquatic pollution.

16.10 REFERENCES

[1] White, K. E., Lee, P. J. and Belcher, A. S. B. (1980). Time-of-travel and its significance. In: Stiff, M. J. (Ed.), *River Pollution Control*, Ellis Horwood Ltd., Chichester. pp. 275–288.

[2] Poels, C. L. M. (1975). Continuous automatic monitoring of surface water with fish. *Wat. Treat. Exam.*, **24**, 46–56.

[3] Van Hoof, F. (1980). Evaluation of an automatic system for detection of of toxic substances in surface water using trout. *Bull. Environm. Contam. Toxicol.*, **25**, 221–225.

[4] Degraeve, G. M. (1982). Avoidance response of Rainbow Trout to phenol. *Prog. Fish-Cult.*, **44**, 82–87.

[5] Morgan, W. S. G., Kühn, P. C., Allais, B. and Willis, G. (1982). An appraisal of the performance of a continuous automatic fish biomonitoring system at an industrial site. *Wat. Sci. Tech.*, **14**, 151–161.

[6] Bull, C. J. and McInerney, J. E. (1974). Behaviour of juvenile Coho Salmon (*Oncorhynchus kisutch*) exposed to Sumithion (Fenitrothion), an organophosphate insecticide. *J. Fish. Res. Board Can.*, **31**, 1867–1872.

[7] Cairns, M. A. and Garton, R. R. (1982). Use of fish ventilation frequency to estimate chronically safe toxicant concentrations. *Transactions of the American Fisheries Society*, **111**, 70–77.

[8] Drummond, R. A., Spoor, W. A., and Olson, G. F. (1973). Some shortterm indicators of sublethal effects of copper on brook trout, *Salvelinus fontinalis*. *J. Fish. Res. Board Can.*, **30**, 698–701.

Detection of Pollution at Drinking Water Intakes

[9] Lunn, C. R., Toews, D. P. and Pree, D. J. (1976). Effects of three pesticides on respiration, coughing and heart rates or rainbow trout (*Salmo gairdneri* Richardson). *Can. J. Zool.*, **54**, 214–219.

[10] Morgan, W. S. G. (1975). Monitoring pesticides by means of changes in electric potential caused by fish opercular rhythms. *Prog. Wat. Tech.*, **7**, 33–40.

[11] Morgan, W. S. G. and Kühn, P. C. (1974). A method to monitor the effects of toxicants upon the breathing rate of Largemouth Bass (*Micropterus salmoides* Lacépède). *Wat. Res.*, **8**, 67–77.

[12] Rice, S. D., Thomas, R. E. and Short, J. W. (1977). Effect of petroleum hydrocarbons on breathing and coughing rates and hydrocarbon uptake-depuration in Pink Salmon fry. In Vernberg, F. J., and A. Calabrese (Eds), *Physiological Responses of Marine Biota to Pollutants*, Academic Press, New York, pp. 259–279.

[13] Sloof, W. (1979). Detection limits of a biological monitoring system based on fish respiration. *Bull. Environm. Contam. Toxicol.*, **23**, 517–523.

[14] Van Der Putte, I., Laurier, M. B. H. M. and Van Eijk, G. L. M. (1982). Respiration and osmoregulation in Rainbow Trout (*Salmo gairdneri*) exposed to hexavalent chromium at different pH values. *Aquatic Toxicology*, **2**, 99–112.

[15] Thompson, K. W., Deaton, M. L., Foutz, R. V., Cairns, J. Jr., and Hendricks, A. C. (1982). Application of time-series intervention analysis to fish ventilatory response data. *Can. J. Fish. Aquat. Sci.*, **39**, 518–521.

[16] Gruber, D. and Cairns, J. Jr. (1981a). Data acquisition and evaluation in biological monitoring systems. *Hydrobiologia*, **83**, 387–393.

[17] Gruber, D. and Cairns, J. Jr. (1981b). Industrial effluent monitoring incorporating a recent automated fish monitoring system. *Water, Air and Soil Pollution*, **15**, 471–481.

[18] Van Der Schalie, W. H., Dickson, K. L., Westlake, G. F. and Cairns, J. Jr. (1979). Fish bioassay monitoring of waste effluents. *Environmental Management*, **3**, 217–235.

[19] Miller, W. F. (1977). *The Development of the WRC Biological Monitor.* Paper 8 at a WRC Seminar Practical Aspects of Water Quality Monitoring Systems, The Water Research Centre, Medmenham.

[20] Solman, A. J. (1983). Monitoring physiological signals from fish. Water Research Centre Technical Report (in preparation).

[21] Evans G. P. and Johnson, D. (1983). Automated detection of pollution at drinking water intakes, using fish. *Proceedings of IWSA conference Security in Water Supply*, Pergamon. (In Press).

[22] Suzuki, S., Satoh, I. and Karube, I. (1982). Recent trends in biosensors in Japan. *Appl. Biochem. Biotechnol.*, **7**, 147–155.

[23] Holland, J. and Green, A. (1975). Development of a gross pollution detector: laboratory studies. *Wat. Treat. Exam.*, **24**, 81–99.

[24] Stroud, K. C. G. and Jones, D. B. (1975). Development of a gross pollution detector: field trials. *Wat. Treat. Exam.*, **24**, 100–119.
[25] Freeman, L. (1983). The sentry that never sleeps. *Wat. Bull.*, **51**, 12–13.
[26] Wallwork, J. F. (1980). Water supply intake protection – River water data collection and pollution monitoring. In: Stiff, M. J. (Ed.), *River Pollution Control*, Ellis Horwood, Ltd., Chichester. pp. 175–187.

Discussion

Chairman: **A. G. SEMPLE**
Secretary, Water Authorities Association

Experience with fish monitors at Welsh Water Authority has shown that distress can be caused by physical factors occurring in water treatment works, for example pump operation, vibration, and background mains and generation noises. The authors were asked if these were seen as problems which could seriously impair the practical application of fish monitors in the field?

Dr G. P. Evans agreed that mains interference can create problems but this has been overcome in this case by means of the signal processing techniques used. Disturbance by vibration and other 'non-polluting' sources have to be eliminated by moving the fish monitor or discount the fish's response during the moments when pumps switched in or out. A microcomputer will be of great value here. **Mr D. Johnson** added that 2000 hours running time with the fish monitor at Medmenham in a fully used laboratory has not caused any distress to the fish.

The authors were asked about monitors of activity other than ventilation rate and whether other fish species have been tested. **Dr Evans** replied that only rainbow trout have been used at Medmenham. Ventilation frequency has been chosen because it gives a large amplitude signal. Heart beat signals are about an order of magnitude less so detection is that much more difficult. Observations of the fish under test showed that ventilation frequency changes were not necessarily associated with swimming activity; indeed ventilation rate increases before activity rate increases.

The timescale of the development of fish monitors and how far away the WRC Mk III monitor is from being commercially available were discussed. **Dr Evans** said that the current WRC development model needs some field testing. A mobile intake protection system (MIPS) has been developed for this purpose. About 1 years' field testing is planned on two MIPS at various field sites. After this period it is hoped to draw up a specification for the batch production of a series of intake protection systems.

Information was requested on what water was used in the laboratory and whether any problems due to high water temperatures occurred. **Mr Johnson** said that experiments were carried out on dechlorinated tap water spiked with the appropriate pollutant. No problems attributable to high temperatures were found. The authors were asked what advantages the Mk III fish monitor had over earlier versions, what is the the optimum of fish required in a monitor and what are the maintenance requirements of the monitor?

Dr Evans cited the following advantages of the Mk III monitor: an earlier response and more sensitive response from opercular rhythm compared with overall activity. The electrode system is also smaller and more efficient than earlier models and noise is reduced. Increasing the number of fish increases the discrimination of the monitor so this is a matter for individual choice, depending on the requirements at the site. On maintenance, a fish will not lose appreciable weight in the monitor for about 2 weeks, so inspection visits may be required every 2 or 3 days with a change of fish every 2 weeks.

Part 5

ENVIRONMENTAL MANAGEMENT – THE PRIORITIES

CHAPTER 17

Costs of environmental quality to the water industry

W. R. HARPER
Thames Water Authority

J. MOSS
Water Research Centre

17.1 INTRODUCTION

This chapter seeks to identify costs associated with water-related environmental quality, and examines the recent economic background against which these costs have been incurred. It then reviews possible changes in cost patterns, reviews their causes and looks at how decisions on environmental quality might be approached.

17.2 ENVIRONMENTAL QUALITY IN WATER-RELATED ACTIVITIES

The range of activities contributing to the maintenance and improvement of environmental quality should be well-known to delegates. However, they are set out below to define our starting point in looking at cost, and in a way which seeks to identify their contribution to environmental quality.

Environmental quality objective	
Environmental conservation	*Activity*
Preservation of ground water levels	Water resource management
Preservation of surface water levels	Water resource management
Preservation of surface water quality	Effluent treatment
	Effluent discharge control
	Sludge disposal control (coastal)
	Incident alleviation
	Quality monitoring
Protection of land amenity	Sludge disposal control
Preservation and restoration of wildilfe and habitats	Land drainage management
	Fisheries management

Health risk minimisation	Activity
Control of hazards in environment	Effluent treatment
	Effluent discharge control
	Control disposal control
	Potable water quality monitoring
Maintenance of drinking water quality	Water treatment

Such listing highlights the major involvement of the water industry with environmental quality in almost all its activities.

17.3 WHAT ARE THE COSTS?

Some costs associated with the activities above are directly identifiable but in other cases costs associated with environmental factors are largely indistinguishable, embedded within some larger body of expenditure. It is quite impossible to quantify the costs borne by all parts of the public and private sectors in maintaining and improving water-related environmental quality. There are very considerable difficulties in attempting to isolate even RWA expenditure in these areas, but some figures are set out below.

RWA capital expenditure by purpose 1983/84
(£M at 1982/83 outturn prices)

	£M
Improve potable supply rivers	2.8
Improve other non-tidal rivers	16.7
Improve estuarial waters and beaches	26.4
Improve coastal waters and beaches	8.8
	54.7
All other purposes	627.2
Total capital expenditure	681.9

Published capital expenditure purpose analysis does not isolate quality improvements that might be associated with water treatment, sewage treatment and sludge treatment. The aggregate figure for 'improvement in levels of service' in 1983/84 is £135M, within which will be elements for quality improvement. It seems likely that expenditure on environmental quality lies within the range £100M p.a. to £150M p.a., perhaps between 15% and 20% of total capital spending.

RWA capital expenditure has, until very recently, formed part of overall public expenditure, and has, as a result, been subject to a substantial real reduction since 1974. There is every reason to believe that quality related expenditure will have suffered particularly badly during this period, since water authorities have given higher priorities to maintenance and delivery of their basic services.

17.4 REVENUE EXPENDITURE

Revenue expenditure on quality-related services is even more difficult to quantify. Even direct operating costs are not readily related to quality parameters, and administration/technical overheads and financing charges (both of which are substantial) cannot be readily associated with operating costs. The table below sets out (imperfectly) to associate elements of revenue expenditure with what we see to be the major quality-related activities.

RWA directly identifiable expenditure on environmental quality
(Direct operating costs only — no overheads or financing charges)

	1975/76 Actual £M	1981/82 Actual £M	% Change	% RPI Change
Drinking Water Quality				
Water treatment	17.4	52.0	199%	
Surface, etc. Water Quality				
Sewage sludge				
Sewage treatment	87.2	157.7	143%	114%
Sludge treatment and disposal		54.6		
Pollution Alleviation and Quality Monitoring	3.1	9.6	210%	

Direct costs in 1981/82 amounted to some £275M. If appropriate overheads and financing charges are taken into account, it is likely that the costs of activities with a strong environmental component are not less than £500M, perhaps some 25% of total RWA expenditure. Operating costs of RWAs do not formally count as public expenditure for the purposes of Government controls, and have not been subject to direct reduction as has been the case with capital expenditure. These costs are, however, subject to very real restraints through the mechanisms of the public acceptability of charges increases and the Government's Performance Aims targets.

17.5 IMPLICATIONS OF RISING REAL COSTS

As is evident from the above, the RWAs bear substantial and rising costs associated with environmental quality. It is also clear that any increases are taking place against a background of severe financial constraint, when relatively low priorities have been attached to quality improvements. We believe this situation is likely to have produced some or all of the following effects:

(i) severe limitation of 'voluntary' improvement to quality standards,

Ch. 17]   Costs of Environmental Quality to the Water Industry   261

(ii) 'crowding out' of other desirable spending,
(iii) increasing tension between public expectations of environmental improvement and resistance of charges increases.

We see in this situation two important management problems. Firstly, that conflicts arise between those having the power to set (or seek to improve) quality standards and those who have the responsibility (including the financial responsibility) for reviewing a wider spectrum of objectives and priorities. Secondly, a demoralising effect on those seeking to constrain costs and improve value for money who see their efforts offset, and more than offset, by influences they regard as beyond their control.

17.6 FUTURE CHANGES IN EXPENDITURE PATTERNS

We foresee future cost changes arising as a result of three main influences:

(i) the increasing recognition of the need for conservation of resources,
(ii) sustained pressure to secure improved efficiency and effectiveness in public authority spending,
(iii) increasing stringency in externally imposed environmental quality standards.

We also see some pressure on costs resulting from increasing concern to preserve the natural environment.

17.6.1 Conservation of resources

Whilst we believe some of the doom-laden projections about the exhaustion of natural resources to be overly alarmist, there seems little doubt that in the long term increasing attention will need to be paid to resource conservation. In the areas we are considering we see the re-use of water and the separation of re-usable materials from sewage to be the subjects of increasing attention. Both situations will almost inevitably require increases in treatment process costs. We would expect to see these increases justified:

(i) either because the costs of alternatives are excessive, or
(ii) because receipts from recovered materials will reduce net costs.

In either event there is likely to be broad coincidence of environmental and economic interest, as increasing scarcity puts pressure on prices and thus precipitates conservation activity in response to changing cost patterns. We can make no sensible attempt to project either the scale or the timing of the cost increases that will arise.

17.6.2 Improving efficiency and effectiveness

The regional water authorities in common with all public (and private) organisations are under pressure to improve efficiency and effectiveness and so reduce costs. We believe that services with a large environmental quality component cannot, and should not, be isolated from these pressures. There is evidence that significant efficiency improvements can be achieved without any controversial interaction with quality standards. These improvements stem from 'good housekeeping' in things like labour productivity, energy usage, etc., but new technology in process management will play a significant part.

Looking at cost-effectiveness in the quality area may prove more controversial, but the benefits may be considerable and rational debate is essential. We quote three examples of areas where the cost-effectiveness approach is being brought to bear:

(i) Looking at discharge consent standards in packages, aimed at optimising overall river quality with least cost. Such an approach would involve managing numbers of discharges to a catchment area so that variations (up and down) contribute overall to river quality, but at the same time minimise investment and operating costs. Such an approach would depart from the current concern with maintaining (and improving) individual consents, and would bring reality to the concept of river basin management.

(ii) Examining the possibility of differentiating winter and summer discharge standards, taking advantage of the varying ability of the receiving waters to accept pollution loads.

(iii) Reviewing practices of disposal of sludge to farmland in view of the increasing costs of conforming to increasing rigorous standards. Disposal to landfill sites, or dumping on sacrificial land, become economically more attractive, and may in turn lead to 'upstream' savings in treatment costs if sludge quality can be relaxed. (We find it ironical that regulations seem likely to change our perception of sludge from something that had a value for farmland, and was often the subject of a charge, to something which has become too costly even to distribute free.)

We would be very disappointed if such measures were seen as attempts to evade environmental standards, since we feel they look beyond the immediate letters of regulations to the broader aims of environmental quality objectives. We believe that the overall impact of such measures on the environment would not be disadvantageous, and that careful and effective use of resources would be encouraged.

17.6.3 Tightening externally imposed standards

We foresee continuing pressures from the EEC, the International Conventions and various environmental groups for increasingly stringent standards, and these are likely to have a substantial impact upon the water industry. It is difficult at present to foresee where these pressures will focus or to quantify their effects with any accuracy in view of the uncertainties involved. Nevertheless, we feel some estimate needs to be made so that the industry can gain a better appreciation of the potential implications. We have therefore attempted to do this for the three major areas involved earlier of drinking water quality, surface water quality and sludge treatment and disposal.

Cost consequences in relation to drinking water quality will result from the implementation of the Drinking Water Directive, particularly the lead and nitrate standards and are estimated to be around £35M p.a. Looking further ahead, there is a possibility of restrictions on the level of certain synthetic organic chemicals in drinking water supplies; this could require the installation of activated carbon treatments, as proposed in the USA, with resulting cost increases of around £25M p.a.

For rivers and tidal waters it would be encouraging to think there will be a gradual increase in quality as called for in the Dangerous Substances Directive, and this would be in line with the water industry's plans. However, economic pressures may make maintenance of the status quo a more realistic aspiration. Again looking further ahead, pressure to restrict the use of the marine environment for waste disposal could have substantial implications, since around 15% of the country's sewage is discharged with minimal treatment to the sea through outfalls. The treatment of these discharges to produce similar effluents to those from inland works would require an increase in costs to the water industry of around £90M p.a. A similar burden would also face manufacturing industry in treating the large quantities of industrial waste currently discharged through sea outfalls.

Sludge disposal authorities have recently assessed the cost implications of the restrictions in the proposed Directive on the Use of Sewage Sludge in Agriculture at around £15M p.a. Certain authorities dispose of their sludge to sea, and restrictions on this outlet could increase their costs by around £30M p.a.

These costs implications for the UK are summarised on the next page.

Not all these costs would fall upon the regional water authorities. Substantial elements would, however, be borne by the RWAs and they would, by comparison with the levels of expenditure outlined earlier, represent very significant increases.

17.6.4 Conservation of the natural environment

Water undertakings, with their close associations with the natural environment, feel particularly the growing pressure for conservation of landscape, habitats, flora and fauna. Costs arise from restrictions on land use, delay in policy implementation, choice of least-cost alternatives and restrictions on operational practice. Measurement of benefits in these areas present profound conceptual and practical problems.

Estimated major costs implications resulting from possible
environmental restrictions (at 1983 price levels)

Source ref.	Area of expenditure	Estimate cost increases		
		Capital £M	Operating £M p.a.	Revenue impact £M p.a.
	Drinking water			
[3]	Lead	400†	7	20
[4]	Nitrate	60	11	16
[5]	Synthetic Organics	100	18	27
	Rivers, estuaries and coastal waters			
[6]	Sea Outfalls (inc. treatment)	600	140	180
	Sewage sludge			
[7]	Utilisation on land	75	10	15
[8]	Disposal to sea	85	25	32

† Including £230M expenditure by houseowners for repalcement of lead service pipes and and storage tanks.

Source references are set out at the end of the chapter.

17.7 APPROACHING DECISIONS ON THE COSTS OF ENVIRONMENTAL QUALITY

We would fully accept that decisions on environmental quality cannot be dicated by cost, but cost cannot be left out of the decision-taking process. In the areas of efficiency, cost-effectiveness and the economics of natural resource conservation, we believe reasonably well-established techniques of financial/economic analysis are available to assist decision-taking. The decisions associated with regulation of quality standards and the actual environment are not readily susceptible to financial analysis.

Techniques of cost-benefit analysis should, in theory, offer assistance, but the problems of practicality and subjectivity (particularly in benefit measurements) can undermine their usefulness. However, we would wish to see the essential elements of an appraisal process employed, that is an attempt made to list objectives, impact, costs and benefit.

The fact that some elements of such a process are not always apparent in considering standards and regulations should be of general concern. At present, a variety of agencies are responsible for setting standards to protect the environment and these will indirectly impose costs upon the water industry. The situation is rapidly changing however and the focus of environmental legislation has already shifted to Brussels. There are four important consequences resulting from this shift:

(i) There is an increasing separation between the authority imposing the standards and the organisations having to meet them.
(ii) There is a trend towards building rigidity into environmental standards, often without any scientific basis.
(iii) The creation of a central point for environmental legislation within the Community has provided a focus for the efforts of pressure groups within Member States and this resulted in such groups assuming an increasingly prominent role in the formulation of environmental policy.
(iv) Political compromise is becoming more noticeable in European environmental legislation, often at the expense of scientific logic.

In our view these changes are not always conductive to bringing forward adequate information on which to found objective decisions. The absence of such information can make difficult the reconciliation required between special interests and the wider interests of the public at large.

17.8 ENVIRONMENTAL QUALITY

We find it difficult to convince ourselves that, in general, the consumer receives good value for money from rigid environmental attitudes and legislation. We believe that for most people the important part of the environment is that with which they have most frequent contact. We therefore believe that most people are willing to pay a 'reasonable' amount for a reliable supply of good quality drinking water, for the satisfactory removal of waste water and its subsequent safe disposal, to prevent contamination of foodstuffs and to prevent pollution or destruction of natural amenities. We recognise that a wide variety of conditions exist in the world, and that significant differences exist even within the developed countries. We feel that the extent to which people are prepared to pay for further improvements in water services depends largely on local (or even personal) preceptions, linked largely to standards of living. This view relies upon the observed willingness and capacity of people to improve 'the quality of life' as personal and national wealth increases. What constitutes good environmental value for money is Surbiton may be regarded differently in Salford, and will certainly be regarded differently in Sao Paulo. We can understand why environmental consequences associated with waste disposal at sea cause more concern in America than in Algeria. As far as the developed countries are concerned, current trends indicate that continued economic growth can no longer be relied upon to sustain environmental quality improvement. Such improvements may need to be considered instead of, rather than in addition to, other desirable social objectives. We are supported in this view by developments in United States policies towards sea disposal where initial proposals for a complete cessation have been modified because of a clearer appreciation of the costs in relation to the resultant environmental benefits. In a perfect world environmental quality may be priceless. In the real world, it is increasingly necessary to know what that price is.

17.9 REFERENCE SOURCES

[1] RWA 1983 summaries prepared by NWC.
[2] RWA Annual Reports and Accounts 1975/76 and 1981/82.
[3] DoE/NWC Standing Technical Advisory Committee on Water Quality Lead in Potable Water Sub-Committee. Report of the Expert Advisory Group on Costs. Lead in Potable Water Technical Note No. 4. June 1981.
[4] Water Research Centre. Nitrate in Drinking Water. Water Research Centre Report 9-M/2. April 1982.
[5] Water Research Centre. Organic Compounds in Drinking Water. An Operational and economic assessment. Water Research Centre Technical Report TR 149. November 1980.
[6] Private communication. (Water Research Centre internal report.)
[7] Private Communication. (DoE/NWC Sub-Committee on the Disposal of Sewage Sludge to land.)
[8] Moss, J., Critchley, R. F. and Lack, T. J. Further Evidence to the Royal Commission on Environmental Pollution concerning the environmental effects of sewage sludge disposal at sea and comparison with alternative options. Water Research Centre Report 375-M/2. August 1982.

CHAPTER 18

The impact on manufacturing industry

T. FARQUHAR
Albright and Wilson Limited

The activities of man on the surface of the Earth are, and always have been, primarily aimed at taking substances from the Earth's crust, manipulating these substances for his sustenance and well-being and then, sooner or later, returning them back to the Earth's crust whence they came. Since the beginning of the industrial revolution over 200 years ago, the scale and complexity of these activities have increased enormously. Technical developments based on scientific research have brought about an explosion in the world's population and has enabled the majority of this increased population to experience a material standard of living which would have been beyond comprehension only two centuries ago.

Notwithstanding scare stories about nuclear weapons ending life on Earth or at least returning a greatly depleted population to a state of nature, the probability is that man will progress in the future to an even higher level of technical development. In so far as this will involve an increase in the rate of manipulation of materials on the Earth's surface, so more care will have to be given to finding a sensible balance between material prosperity on the one hand and the benefit of the natural environment and its preservation for the future on the other.

It started to be realised just over 100 years ago, that there is a necessity to intervene deliberately in the process of change so as to avoid adopting easy short-term expedience to the detriment of the long term. The expansion of the chemical industry in Germany and the United Kingdom during the late eighteenth and early nineteenth centuries, was looked at as of such great benefit at the time that few people recognised the damage being done to our environment as a result of its operations. This period also saw the beginning of the coal gas industry with all the benefits that provided. It also saw the by-products of the gas industry, ammonia, tars, aromatic hydrocarbons and sulphur compounds, being poured into rivers with devastating consequences.

Earlier in the nineteenth century, concurrent with the beginnings of the chemical and gas industries, another development took place which would certainly be judged today as an important development in protecting the environment. This was the introduction into London and other major towns of the water-carriage system for sewage disposal. Before this time, sewage and wash waters of all kinds were merely thrown into the streets so contributing largely to the outbreaks of plague and epidemics which raged from time to time. Here again, the undoubted advantage in cleaning up the streets and reducing the spread of disease had its disadvantages. All the newly built sewers did was to transport human and other wastes to the rivers which, during the nineteenth century, degenerated into what Dr William Budd in 1859 described as 'Seething and fermenting cloacas with stench so foul as had never before ascended to pollute this lower air'.

Even in more recent times, the rise in living standards has been demonstrated to be causing environmental problems. Until the 1960s, London and other major cities suffered on a number of days each year from a serious level of air pollution. Growing populations in our urban areas coupled with the use of increasing quantities of fuel for comfort heating in homes, resulted at times in a build up of sulphur dioxide and particulates in the atmosphere which had a significant impact on the health of the population. In December 1952, the great London smog which reduced visibility to almost zero for a week was statistically shown to have killed over 4000 people who would not otherwise have been expected to die at that time.

Although the realisation that industrial development and the use of the products of industry by a growing world population, carry the need for environmental control, it is possible that history will relate that it was only in the 1970s that most advanced countries fully accepted the fact. Prior to the 1970s, legislation to protect the environment was often not comprehensive and had been confined to the imposition of controls in those areas where environmental damage was most obvious. History should also relate, however, that in spite of this piecemeal approach, efforts towards the maintenance of a clean and healthy environment did start during the latter half of the nineteenth century. Notwithstanding the popular belief put out by those who seek to make protection of the natural environment a political issue, man in 1983 is not polluting the environment more than ever before. Indeed, the very advances in scientific thinking which have brought about the jump in material prosperity, have also brought about the realisation that man is part of a complex natural system which must be protected. Scientific development has given man the ability to examine and understand his surroundings as never before and to provide the means by which he can enjoy to a large extent, the best of both worlds, material prosperity and a clean environment.

Nobody should be in any doubt that maintenance of a clean environment costs money. There are those who would deny this citing for example the fact that in some locations, potable water is more expensive than it would otherwise be if the local river was not polluted by upstream discharges. It is probably correct that in choosing between the various options and seeking to maintain the most efficient industry, while protecting the environment to the maximum degree, the right solutions are not always made. One factor which has come into the equation in recent years is that in many instances we are not now looking at a badly polluted environment and wondering what we can afford to spend to improve it nearer to an acceptable quality. In many cases, the point has now been reached where it is arguable how much further it is necessary to go. The fact that so much has been done at such considerable cost makes it more and more imperative that future expenditure should be focused sharply on those areas of the environment where improvement is still shown to be necessary. The momentum of the past must not be allowed to sweep us beyond, what in years gone by, would have been looked on as the ultimate target.

Accepting that a balance must be struck between pure materialism and protection of the environment, decisions will have to be taken at the political level on how it will be managed. Ultimately of course any increase in industrial costs required to provide for a clean environment will be carried by the final customer, the man and woman in the street. The point in the economic chain at which the cost of protecting the environment will be extracted must however be decided. It may be tempting for some politicians to suggest, particularly in times of high unemployment and industrial failures, that some of the costs of protecting the environment should not be imposed directly on industry but should be carried by public funds or in other words, by the taxpayer. This it may be argued will maintain the prosperity of industry and will spread the cost of protecting the environment thinly throughout the economic system. The flaw in this argument and the reason why it enjoys little support, is that it provides no incentive to industry to locate its factories or design its processes in a way which will raise the least environmental difficulty. The well-known principle that the polluter pays, meaning that the cost of mitigating the results of pollution and also the costs of preventing pollution should be carried by those directly responsible, comes much closer to a rational system and is widely supported. Unfortunately, like many slick definitions which sound clear and simple, the principle that the polluter pays can in practice mean all things to all men.

The exact level of expenditure by industry under the heading of protecting the environment is not known with any precision as is demonstrated by the fact that attempts at assessment by different authorities come out with widely different answers. The problem arises from the fact that most calculations have been based on the assumption that examination of the process used in a production plant will make it obvious which parts of it are specifically there to protect the environment and which are not. A typical example from the chemical industry demonstrates the problem.

Substance A is converted by oxidation with air to substance B which is a gas at the reaction temperature and is absorbed from the process train in a water scrubber to give the required product in liquid form. The effluent gases from the scrubber which must at the very least contain the nitrogen from the original air are discharged from the scrubber to atmosphere. In practice, this discharge will be liable to contain in addition, firstly, a proportion of substance A which has failed to be converted to B in the first stage and has not been removed in the scrubber, secondly, a proportion of substance B which has not been removed by the scrubber and thirdly, a proportion of other substances some of which may have been present as impurities in A and some of which may have been formed due to extraneous side reactions within the process itself. It may be assumed that some or all of the substances other than nitrogen in the effluent gases are toxic, noxious or polluting and so controls are imposed on the concentrations and/or quantities of these materials in the discharge to the atmosphere.

In the case of an existing plant operating these processes which before the end of its useful life is found to be causing unacceptable atmospheric pollution, the remedy might be to construct an 'add on' treatment plant at the end of the train and so reduce the level of pollutants in the discharge. If the materials so removed can be collected and sold, then the cost of this 'add on' treatment plant may be considered to be partially environmentally and partially productive. The problem is the two proportions will be liable to vary very widely due to short term changes in economic circumstances. The proportion of the capital which is unjustified in relation to the additional product obtained, may be considered environmental but its calculation will depend entirely on an assessment of the justified proportion and this will vary in terms of the return which the owners of the plant believe they should be earning on their capital. This latter factor will depend to a considerable extent on the state of the proprietor's finances, to the rate at which he can borrow money, to the other demands being made on his resources and to the return which he believes he could make by investing the money elsewhere. None of these variables is related in any way to the nature of the process or the pollutants discharged from it.

Other problems would face the unwary in assessing the environmental cost of such a process. The cost of purchasing substance A may increase giving rise to a corresponding increase in the selling price of the final product. Provided other costs do not change at the same time, it will obviously pay the processor to operate at a higher materials efficiency than before even if this involves additional capital expenditure to allow him to do so. A substantial increase in the cost of purchasing A may even result in the cost of the 'add on' treatment plant, originally put in largely for environmental purposes, becoming fully justified in economic terms so that the environmental component in its cost could fall to zero over night. Basing calculations on environmental cost on parameters which are subject to so many short term variables and thus providing widely fluctuating answers, is hardly a sound basis for arguing and deciding

Ch. 18]   **The Impact on Manufacturing Industry**   271

where the balance of interest between material prosperity and environmental protection should lie.

Obviously such arguments can be taken further into even more complex cases. When hitherto uneconomic processes are adopted because they are environmentally clean and so can begin to compete with potentially polluting processes made uneconomic by heavy environmental factors, comparisons become almost impossible. A close examination of such a new clean process may reveal no expenditure specifically aimed at protection of the environment but it would be misleading to conclude that none was present. The former dirty process would have been capable of supplying the same product more cheaply but for environmental factors and so a proportion of the selling price of the product from the new plant is in reality being used to protect the environment. Even if the assessor is aware of the existence of this factor, and tries to work out the hypothetical cost of operating the old style process, he is likely to have to make so many guesses as to make the exercise almost meaningless. If the starting point of the former process was a mineral or other substance which has now been taken off the market due to lack of demand, it may be impossible to obtain a meaningful quotation for its supply.

Such examples demonstrate the extreme difficulty of making any assessment of how much it is costing industry to protect the environment. Nevertheless, even a rough qualitative assessment can leave no doubt that the sums are considerable. Attempts by a number of representative bodies including the European Commission to obtain data on environmental costs, have usually been based on the issuing of questionnaires to be filled up by selected industrialists thus side-stepping the difficulties. It seems doubtful if the results of such exercises add greatly to knowledge of the subject.

What is needed if any meaningful progress is to be made is for codes of practice to be agreed with the accountancy profession so that assessments can be made using standard procedures. Such codes would have to be based on acceptance of the fact that there must be a balance between desirable simplicity and the extreme complexity which would be inevitable if absolute logicality was attempted.

The Du Pont Company in the United States is among those who have attempted such an exercise. Basic information on the cost projections was obtained by a plant-by-plant survey conducted by a specially appointed team of accountants and engineers. A ten-year period was selected from 1975 to 1985. Not only did the Du Pont team make an assessment of the environmental costs, they also attempted to assess those costs which they considered to be excessive on the grounds that they gave no identifiable environmental benefit in terms of human health, wild life protection, vegetation, conservation of resources or recreational advantage. Here, in brief, is what the survey revealed.

Thirty per cent of Du Pont's projected capital expenditure of $10b during the period that is $3b would be required for pollution, including noise control and would give no direct return to the company. This $3b would not therefore be available for investment in productive capacity. Of the $3b 75 per cent would yield no practical benefit to the environment. It was estimated in addition that by 1985, $1b of Du Pont's operating funds would go each year to meet environmental requirements. It is of course true that the Du Pont results and in particular the proportion of its environmental expenditure which was considered gave no environmental improvement, was conducted against a background of conditions in the United States. Such an assessment might suggest a somewhat lower level of unnecessary expenditure in most European countries, although an estimate made by one of the German chemical majors 3 years ago, suggested that one-third of all capital expenditure was required solely to protect the environment. It is unclear however if this work followed the same accountancy rules as were used at Du Pont.

The conclusion therefore that must be drawn from what is known of environmental costs in industry is that it is impossible to assess with any precision what is certainly highly significant in terms of national material prosperity. When people are demonstrated to be dying in statistically significant numbers as a result of an industrial operation, it is clearly necessary to do something about it even if it does mean a reduced material standard of living for the majority. As time goes on, the decisions are becoming more and more difficult to make. Continuing improvements on environmental control become increasingly expensive in relation to the inprovements obtained. There is now no evidence from mortality statistics that the air people breathe in London is having a material impact on the health of the population compared with those living outside the metropolis. It may be that people living in London are still less healthy to a degree which is too small to show up in the statistics. How can one assess what material sacrifices should continue to be made to improve London's air quality further 'just in case'? Decisions on such matters must become more and more subjective but it is important for the politicians who will ultimately make them to be made aware of the implications in so far as they are known. Is it better to afford to have roast beef for dinner every Sunday and live beside a river which sometimes smells in summer or is it better to eat sausage and mash every second Sunday instead and to live beside a river which never smells at all? It is all a matter of politics!

CHAPTER 19

Environmental protection – what is it worth?

GEOFFREY LEAN
The Observer

It is part of a journalist's lot to ask impossibly simple questions – and perhaps no group of people are put to more trouble by this irritating habit than environmental scientists. You may have spent years on research, carefully evaluated the implications of your findings, and then finally published a paper in which every word has been weighed to the nearest nanogram, and which seems masterfully clear and comprehensive to you. And then some benighted journalist comes on the phone with questions like: 'What does all this really mean to our readers?'; 'How many people are going to die from this hazard you have been working on?'; and 'How much is it going to cost to clean it up?'

Anyone here who has ever suffered in this way is going to enjoy the next twenty minutes – for now the boot is on the other foot. The title of this chapter must surely pose the most impossibly simple question of them all.

Obviously, virtually everybody believes in protecting the environment and virtually nobody thinks it should be preserved at any cost. There are, of course, a few exceptions – and in my job you tend to meet both the hardliners who argue that pollution is good for you, and the environmental extremists who say that growth should cease and people even be allowed to starve so as to safeguard nature. The rest of us try to decide on the balance between pollution and protection, destruction and preservation by weighing the costs and benfits of different types of action – or inaction. The trouble is that this is extremely hard to do.

The costs of environmental protection are difficult to assess accurately – and the benefits (or costs of inaction) are almost impossible to quantify. So cost–benefit analysis has its dangers. While it is the best crude tool we have for the purpose, and can indeed provide valuable perspective, it can also distort the decision-making process.

The costs of environmental protection are the easier to perceive, partly because those who will have to bear the costs usually take pains to make their case known. Unfortunately, Governments — and, indeed, all of us — are often disproportionately dependent on the estimates of costs provided by the industries and other bodies who will have to bear them — and it is often in their interest to exaggerate them.

Thus German petrol companies in the early 1970s countered pressure to reduce the maximum level of lead tp 0.15 grams per litre by claiming that they would have to undertake capital investment amounting to 1100 million Deutschmarks in their refineries, cars would have to be extensively modified and the cost of fuel would rise by 25p a gallon if this target were to be achieved. In the event their capital investment was just over a quarter of their estimate (at 277.5 million Deutschmarks), no changes needed to be made to cars and the price of petrol remained the same.

Similar arguments were heard during the recent debate over lead in petrol in Britain. The Associated Octel Company Ltd, makers of lead additives for fuel, stated in a booklet designed for public consumption: 'Legislation (to completely remove lead from petrol) would immediately be felt by the British motor industry who would have to redesign and retool to produce engines suitable for lead-free petrol. Even if the industry was capable of dealing with such a change and able to find the investment needed, the individual and company purchaser would notice a difference in showroom prices and very significant adverse differences in fuel economy and performance. There would be a considerable increase in the cost of petrol due to increased refinery costs with all that implies for the average 12 000 miles a year motorist and for the average 50 000 miles a year company salesman. Our oil imports would dramatically rise and the eventual cost would exceed £450 million annually. Without the addition of lead, Britain would pay a very high price for motoring' [1].

It sounds compelling enough and it was not without influence. Both civil servants and ministers during the controversy would privately emphasise that they were almost entirely dependent upon industry's estimates of the costs of change. As it happened, however, the Royal Commission on Environmental Pollution was making an independent study of the issue, and came out with different figures [2]. It calculated that moving from 0.15 grams of lead per litre to lead-free petrol would probably only require £70 million in extra crude oil, that car prices would probably rise by a figure in the order of £6–£14 a vehicle, that unleaded petrol would probably cost ' a few pence at the most' per gallon more than leaded fuel, and that the cost to the average private motorist driving 10 000 miles a year would be about £10. Furthermore they added, even expressing costs in this way was misleading, because over the next years they would be far outweighed by improvements in car design which would greatly increase fuel economy. So the average motorist would not suffer a penalty, even a small one, but forgo a small part of some future benefit. The Government found this argument persuasive and decided to work for a ban.

Ch. 19]   **Environmental Protection — What Is It Worth?**   275

To take another example from a similar field; in June 1983 the Wood Inquiry reported [3] on proposals to ban lorries from London. In evidence the road haulage industry estimated that if all lorries over 16 tons were allowed to proceed no further than the M25 and had to transship their loads into smaller vehicles for delivery in London it would have to incur massive extra running costs; its estimates ranged from £250 million a year to £750 million. Again, to the layman this looks not unreasonable. The committee, which included academics, industry representatives and environmentalists tested this in a major exercise, using a computer model agreed by all its members. This came up with the result that there would actually be *a saving* to the industry of £12 million a year. At this point the two industry representatives (from the haulage and retail industries) dissented. They produced their own estimate of a cost of £73 million — still very much less than the original figures.

I have taken examples from similar areas, but others could be given. The asbestos industry would provide one example — with the difference that, by and large, regulatory authorities have in the past placed overdue weight on the industry's own estimates of costs, with the results that controls have not been as tight as they might have been.

One of the most interesting aspects of these three examples is that events indicated not only that the massive costs originally estimated could not be justified, but they were either outweighed by benefits (when these were calculated) or that they were so small that any benefits would almost certainly swamp them.

Interestingly, this seems to hold true even when trying to assess the costs and benefits of one of the most expensive pollution control enterprises of all, the reduction of emissions of sulphur oxides that, *inter alia*, give rise to acid rain. In 1981 The Organisation for Economic Co-operation and Development made a first, rough, attempt to quantify them [4]. The results must be taken with caution as the report was designed to present a methodology for a cost-benefit analysis, rather than produce definitive results — but they are so striking as to be worth noting. As expected, the costs were enormous, $2.7 billion a year for Western Europe to cut its emissions by 37 per cent from 1974 levels by 1985. Yet the benefits in better health, diminished corrosion and better crops ranged from $1.8 billion a year to $15.6 billion (the range mainly reflects uncertainties over the dose—response relationship between sulphur and ill-health).

Significantly, the study says that it almost certainly underestimated the benefits. It only considered the effects of corrosion on steel and did not take into account the damage done to stonework and other building fabric — let alone attempt to quantify the loss of mankind caused by acid-induced damage to ancient monuments like the Parthenon and Trajan's column in Rome. It only considered the damage to *agricultural* crops, omitting forestry — though this is of major concern to countries like Sweden and West Germany. Also it confined its assessment of the health costs of the pollution to treatment costs and days lost through morbidity — making no attempt to compute a mortality cost.

This raises what some consider to be the major failing of the whole cost – benefit approach — the virtual impossibility of ascertaining a true cost for the effects of pollution and hence of the benefits of reducing it Does adding up the cost of treatment and the amount of income lost really provide an accurate assessment of the real cost of illness to an individual and his or her family? If it is difficult to arrive at a true cost for morbidity, how can one for mortality ever be assessed? How can any analysis measure the cost of the ruining of some of the world's greatest cultural treasures? And, if this is difficult, how much harder is it to cost the worth of clean air and water, beautiful countryside, or even the meanest endangered species?

Fritz Schumacher used to argue that even to attempt to do this was wrong. 'It is a procedure' he said 'by which the higher is reduced to the lower and the priceless is given a price. It can therefore never serve to clarify the situation and lead to an enlightened decision. All it can do is to lead to self-deception or the deception of others; for to undertake to measure the immeasurable is absurd and constitutes but an elaborate method of moving from preconceived notions to foregone conclusions. The logical absurdity, however, is not the greatest fault of the undertaking; what is worse, and destructive of civilisation, is the pretence that everything has a price or, in other words, that money is the highest of all values. If the economist remains unaware of the fact that there are boundaries to the applicability of the economic calculus, he is likely to fall into a similar kind of error to that of certain medieval theologians who tried to settle questions of physics by means of biblical quotations' [5].

Whether or not one accepts his moral strictures, it is hard to disagree with his conclusion that putting a price on the priceless is impossible. The difficulty of trying to assess the benefits of environmental protection is compounded even further by our inescapable inablity to foresee exactly what the effects of environmental degradation may be.

Recently a new species of corn (*Zea diploperennis*) was found in Mexico. It is apparently immune to at least three maize virus diseases for which there is no other source of immunity and so offers a unique chance to plant breeders to produce protected strains. It is a perennial, and if this attribute were to be cross-bred into current commercial forms of annual corn it could eventually reduce the massive cost of season-by-season ploughing and sowing. Furthermore it can grow in wet soil, an attribute that could be used significantly to increase the acreage that could be planted with corn worldwide. The discovery was only just made in time; for the corn was found growing on just a few acres of common land marked for development.

Clearly the benefits to be gained from the corn will outweigh many million-fold the small cost of failing to develop the land where it grew until its seeds were collected and it was successfully cultivated elsewhere. (Leaf blight eliminated 15 per cent of the entire US corn crop in 1970, at a cost of $2 billion, until it was halted with the aid of a blight-resistant strain whose genetic material also originated in Mexico.) Yet no-one could have known this in advance. The people who wanted to develop the land and the authorities who permitted them to do it, could not possibly have known what the cost of their action would have been.

This story happens to have a happy ending; but a wildlife species is already disappearing every day and the rate is accelerating. By some estimates one million species may have become extinct by the year 2000 [6]. Yet we have not even discovered more than about 20 per cent of the world's species and only studied a small fraction of those we have identified [7]. A species due to become extinct tomorrow may contain the attributes needed to save a vital food crop from some future plague or to cure cancer. (The rosy periwinkle (*vinca rosa*) for example, has enabled us to quadruple the chances of successfully treating Hodgkin's disease, given lymphatic leukemia patients an excellent chance of recovery and proved effective in fighting other cancers.) But even if a cost–benefit analysis were made of whether to flatten the particular piece of tropical rain forest in which the last of this unknown species lives, there is no way it could take into account the magnitude of the benefits about to be irretrievably lost.

Uncertainty about the health effects of pollutants presents a similar problem. It is, for example, by no means certain that low levels of lead do diminish children's intelligence and the Royal Commission took great care to skirt around this issue. It will probably never be proved, for as Sir Henry Yellowlees, then Chief Medical Officer at the Department of Health and Social Security put it, in 1981 'we are dealing here with the biological sciences where truly conclusive evidence may be unobtainable'. What goes for lead goes too for many other pollutants.

Such uncertainty is, of course, made even greater by the long time scale over which cancers and some other diseases manifest themselves. This means that even if we are alert and lucky to spot a new outbreak of disease and even if we can ascertain its causes, we will only be able to do so several decades after the decision to release the responsible substance has been taken and the damage done.

Similar time-lags affect the possible repercussions of pollution on the natural environment. If it is true that our increasing burning of fossil fuels is warming the atmosphere to the level that the seas may rise and inundate coastal cities, or — more probably — cause extensive shifts in the pattern of rainfall and crop producing areas (so that the Soviet Union rather than the United States might, in the view of some recent projections become the world's main grain growing area), then the costs of letting carbon dioxide levels in the atmosphere rise must enormously outweigh the costs of rigorous programmes of energy conservation and development of alternative forms of energy. The trouble is that we will not know whether the forecasts are correct or not until they are fulfilled, whenever and if ever that is, by which time it will be far too late.

Some areas seem to be affected by both time-lags in health and time-lags in the natural environment and, if I may stray into an area where I am the least expert person in the room, this appears to be the situation as regard nitrates, at least in groundwater. If it is true that a nitrate front is working down towards main aquifers, and if the present debatable evidence that it may start a chain of events in the body that lead to cancer turns out to be correct, then we may regret failing to take adequate action now.

All the factors I have mentioned so far tend to bias the process of decision making against taking measures to protect the environment. The decision maker is in a difficult position. On the one hand there are parties who can vigorously point out what environmental measures will cost them. They may, or may not, exaggerate those costs as some have in the past, but, at the very least, they can make a strong case for their point of view. Those who would benefit from the measures are usually far less well organised and have a much less clear case to put.

Beneficiaries are often a much less coherent and concentrated group of people than those who would have to pay the costs. May I put on my other hat at *The Observer*, that of an economic writer, for a moment, to illustrate this? Almost every study shows that protectionism generally does more damage than good. It sends up prices to consumers and it causes more jobs to be lost in other sectors of the economy than it saves in the industries protected. But the advocates of protectionism are organised, sectoral interests, whereas those who would lose as a result of it almost inevitably are not. The management and unions from the textile industry, for example, well know that without protectionism they will lose jobs and business and can present facts and figures, sometimes with exaggeration, to Government. The industries and consumers who will suffer do not know what the effects on them will be and so will never exercise a countervailing pressure. And so successive Governments, even ones with rigorous ideologies, end up by bowing to the sectoral interests.

Even when the beneficiaries from environmental protection are aware of their situation, they will usually have a less compelling-sounding case to put. The benefits that would result are often unprovable, often unknown, usually unquantifiable and usually take time to materialise. The decision maker faced with an obvious cost that will anger a vocal group of people at once and benefits which are uncertain and unquantifiable, which take effect long after he has left office and which the beneficiaries may never even be aware of — let alone trace back to him — is understandably likely to lean in favour of inaction.

Of course, despite this, action to protect the environment is often taken, and I mentioned at the beginning of this paper the action both Britain and West Germany have taken over lead in petrol. Normally the action has only followed public pressure, rather than been taken solely on the basis of a cost—benefit analysis. This, I believe, underlines rather than undermines my point, that the pressures for inaction will generally succeed unless there is an overriding countervailing force — which, usually, as far as politicians are concerned, will come from public opinion.

For in the end, the only answer to the question 'Environmental Protection — What is it worth?' will be provided by the public. For all practical purposes, it will be deemed to be worth what public opinion thinks it is worth. Such indications as we have, suggest that the public evaluate it very highly — much more so than might be supposed.

Newspapers are only just beginning to catch on (and I believe *The Observer* has been in the van) to the interest people take in the environment. Many more people belong to the most populous conservation groups than to all the political parties combined. Many more watch *Wildlife on One* than *Match of the Day*. A recent public opinion poll showed that one-fifth of the electorate would 'seriously consider' switching their vote to a main political party that committed itself to reduce wastage of natural resources; that over half would happily pay a penny in a pound more on their income tax to reduce such wastage and protect wildlife and the environment and; that over half said attractive countryside and an unpolluted environment made 'a valuable contribution to their lives' as compared to a third who listed access to a car or sports facilities, and less than fifty who mentioned the theatre [9].

The effect of public opinion has been more and more evident in environmental decisions, most recently in the British decision to ban lead in petrol and in the international resolution to stop all commercial whaling. Of course, public opinion is not always strictly rational and may force through decisions that are hard to justify scientifically. The ban on commercial whaling is just one example among many. There was overwhelming justification for some of the earlier selective bans on hunting the most endangered whale species — but no strong scientific evidence for banning all commercial whaling; Minke Whales, the main species still hunted, remain plentiful and in no danger of extinction. Public opinion, however, here and in most countries, was overwhelmingly in favour of saving all whales.

What happened, of course, was that people made a value judgement, rather than a strictly scientific one. They decided that whaling was a cruel and unworthy occupation. The strength of this view is, in itself, an interesting example of the interest in environmental issues. Here are a family of animals which most people have never seen and will never see and which have none of the cuddly appeal of baby seals — and yet which attracted enormous public sympathy. Without doubt most of the support for banning whaling was emotional. Scientists and economists — and even environmental journalists — distrust emotion, because it does not, and should not, be allowed to distort our analyses and judgements. But it is a perfectly valid element in public opinion. In the end, most environmental decisions will have to be taken on the basis of popular value-judgements and emotion, as well as hard fact and cold-blooded cost—benefit analysis, plays its part in making them. And when the facts are unclear and cost—benefit analysis is often incapable of providing a balanced judgement, the injection of a good dose of public gut-feeling into the decision-making process is inevitable.

REFERENCES

[1] Associated Octel (1981) *Lead. Do you know enough about the Lead controversy?*
[2] Royal Commission on Environmental Pollution; Ninth Report; *Lead in the Environment*, HMSO, April 1983.
[3] Wood, D., *et al.*, (1983) *Report of the Independent panel of Inquiry into the Effects of Bans on Heavy Lorries within London.* GLC, June.
[4] Organisation for Economic Co-operation and Development (1981) *The Costs and Benefits of Sulphur Oxide Control*, OECD, Paris.
[5] Schumacher, E. F. (1973) *Small is Beautiful, A Study of Economics as if People mattered*, Blond and Briggs.
[6] Myers, N. (1979) *The Sinking Ark*, Rockefeller Brothers Fund, New York.
[7] United Nations Environment Programme, World Wildlife Fund, International Union for the Conservation of Nature and Natural Resources, *World Conservation Strategy*, 1980.
[8] Sir Henry Yellowlees, Letter dated March 6th 1981, to Sir James Hamilton, Permanent Secretary, Department of Education and Science (published in *The Times*, February 8, 1982.
[9] MORI, *Poll on Resource Use in Britain*, published by the Conservation and Development Programme for the UK, July 1983.

Discussion

Chairman: **R. T. WHITELEY**
Chairman, Water Research Centre

The discussion was opened by **Mr W. N. Richards** (Strathclyde RC) who spoke about concentrations of lead in water in Glasgow and Ayr. In Glasgow the drinking water pH is 6.3 and prior to any action to solve the problem, more than 50 per cent of the samples did not comply with the EEC standard ($100\,\mu g/l$) and more than 5 per cent exceeded $1000\,\mu g$ pb/l; In Ayr, before 1980, the pH was 5.5, more than 70 per cent of samples exceeded the EEC standard and more than 14 per cent exceeded $1000\,\mu g/l$ (some reaching 5 mg/l). The water was treated by lime addition, raising the pH to between 8 and 9 and in 1980 in Glasgow more than 95 per cent of samples were less than $100\,\mu g/l$ and around 90 per cent of samples in Ayr currently are less than the standard. The costs involved for Glasgow water were about £318,000 (1978) for capital and current revenue expenditure is £48–50 000. For Ayr the corresponding costs were £15 300 capital (1981) and £6800 revenue costs. Compliance is still not totally satisfactory and the water could be treated further by adding ortho-phosphate. This could reduce the lead concentrations by a further 50 per cent but the additional costs would be £200 000 (capital) and £225 000 (revenue) for Glasgow. For Ayr the corresponding costs would be an additional £18 000 (capital) and £18 000 (revenue). In Glasgow in 1977 the mean lead level in the blood of women was about $14.6\,\mu g/100$ ml and in Ayr in 1980 the level was $22.9\,\mu g/100$ ml. There has subsequently been a 40 to 45 per cent reduction in blood-lead levels and both areas would now pass the EEC blood-lead screening programme. Therefore is it worthwhile to spend considerably more money on further reducing lead levels in drinking water?

Geoffrey Lean commended the efforts to reduce the blood-lead levels but was still concerned about the levels in young children and the possible damage to their brains. The cost of reduced IQ in children might be considerably more than the cost of treating the water further. If, however, the levels were satisfactory then there appeared to be little justification for more spending.

Mr David Walker (National Water Council) emphasised the view presented by Mr Harper and Dr Moss that we not only have to make a choice between environmental improvements we also have to choose between environmental improvements and other desirable social measures. This is an area where journalists can help because they can exert great influences on the population. Journalists should be prepared to discuss the costs and the benefits in the way Mr Richards has demonstrated. So we must co-operate with technical journalists and ask them to sacrifice the quick sensational story in order to spend more time in explaining the difficult choices that have to be made between various ways of improving our environment.

Mr Lean responded by saying that he agreed that cost−benefit analysis of environmental improvement options was the best way (although imperfect) of assessing priorities. However, he made the point that governments may decide priorities on the basis of public opinion and that may be reached without having access to all the available facts. There followed considerable discussion on the relationship between the press and the water industry. Mr Lean quoted examples of being unable to find willing spokesmen to advise him on the nitrates problem. Other speakers drew attention to the distrust of scientists in industry towards journalists who might misrepresent or inaccurately report their statements. The general point was made that only reputable journalists survive and reputation is founded on a mutual trust that facts and opinions given in good faith will be accurately and fairly represented in the press.

CHAPTER 20

Protection of the aquatic environment – research needs

S. C. WARREN
Water Research Centre

20.1 RESEARCH AND ENVIRONMENTAL POLICY

The title of this chapter 'Research Needs' prompts the question 'needed by whom, and for what purpose?'

Since government is the largest single spender on environmental research, environmental policy is a suitable context in which to examine the requirements for research.

Environmental protection is a call on national resources and expenditure and whether the money is collected in the form of taxes, rates, water rates, or as a part of the price of goods, the cost is eventually borne by the individual members of society. At its most fundamental, therefore, a national policy must be the framework within which to make decisions on environmental expenditure.

The relationship between environmental policy and research resembles that between the chicken and the egg. Environmental policy needs to be based on scientific and technical facts which are generated by research, while research looks to policy for guidance on research priorities.

Much of the pressure for environmental protection comes from Brussels and the European Commission appears to be a principal driving force in environmental legislation enacted in the member states. On other environmental matters government seems to find itself responding to a series of threats as and when they occur. There is no obvious plan for the development of environmental policy and much research funding reflects an *ad hoc* response to environmental problems. The absence of long-term policy, which would allow research to be set clear objectives, handicaps scientists in formulating research programmes and allows publicity to exert an undue influence on policy making.

Despite the hope of the European Commission that economic factors should not be allowed to impede environmental improvements, in practice they will. Environmental protection nearly always costs money and it is essential that the available money is spent where it will do most good. Spending priorities at present, however, do not always seem to be strongly influenced by an objective assessment of the cost and benefits of different courses of action.

There are three main sources of anxiety — chemicals, natural wastes and exploitation of resources. The concern relates to possible risks to human health, damage to the ecological balance and damage to its aesthetic properties. In recent years concern has extended to the marine environment. Some of the current problems are shown in Table 20.1 as they affect different aspects of the environment.

Table 20.1
Potential problem areas

| Risk of damage | | Industrial chemicals | Natural wastes | Exploitation |
To	From			
Human health	Fresh water	Water treatment Water re-use Agricultural chemicals	Water re-use	
	Sea water	Industrial discharges	Coastal sewage discharges	
Ecology	Fresh water	Industrial discharges Waste disposal Acid rain	Sewage effluents Animal wastes	Flood relief and land drainage Over-fishing
	Sea water	Dumping at sea of: Industrial and radio-active wastes Dredging spoil Sewage sludge	Sewage sludge dumping Coastal sewage and industrial discharges	Coastal and estuarial developments Over-fishing
Aesthetic aspects	Fresh water	Acid rain	Sewage effluent discharges	Flood relief and land drainage
	Sea water	Coastal industrial waste discharges	Coastal sewage discharges	Coastal and estuarial developments

20.2 RESEARCH NEEDS

20.2.1 Industrial chemicals in the environment

Chemicals enter the environment by a wide variety of routes and in a number of different chemical forms. The traditional UK policy has been to require that the best practicable means should be used to minimise the environmental impact of chemicals. Similarly, the requirement for drinking water is simply that it should be 'wholesome'. European policy, on the other hand, is to set limits to the concentrations which are permitted in different circumstances for a very large number of chemicals.

Unfortunately sufficient scientific evidence is not always available to allow such limits to be set with confidence. There is an understandable tendency to err on the side of caution and this has led in some instances to extreme, even absurd, limits, especially for substances for which proper data is almost non-existent. Furthermore, the determination of the substances involved, at the low concentrations specified, is often very difficult and intercomparability studies commonly reveal wide variations in results between different laboratories. Unless steps are taken to ensure that results of adequate accuracy are achieved in all laboratories concerned, the fairness and practicability of standards may be severely prejudiced.

One consequence of excessively stringent standards is greatly increased costs for industry which may be unnecessary if such low limits prove not to be justified. It is a matter of urgency, for Europe as a whole, that research be undertaken to provide proper data on which to determine maximum allowable concentrations, particularly for some of the more contentious substances. The information required and the consequent research needs are shown in Table 20.2. The research needs can be illustrated by considering specific chemicals which from time to time have received attention.

20.2.1.1 *Heavy metals*

It is easy to demonstrate the toxicity of heavy metals towards plants and animals under laboratory conditions, but unequivocal evidence of effects in the natural environment is harder to find.

Evidence already exists that the biological effects, transport and fate of metals are strongly influenced by their oxidation state, and by the way in which they are combined with other substances. Heavy metals are strongly inclined to form complex compounds and to bind strongly to clays and organic matter. The extent to which they are available to plants and animals in this combined state and, if so, whether their toxic effects are reduced by chemical combination, is much less well understood. The fact that the toxicity of a particular element can be demonstrated in the laboratory need not imply that its presence in the environment is dangerous, although it may be potentially harmful. A better understanding is required of the forms of trace elements present, of the effects of chemical state on bioavailability and toxicity, of the transformation of these forms in the natural environment and of their transmission through the food chain.

Table 20.2
Research needs for chemical pollutants

Requirement for information	Research needs
1. Present position	
(a) How much of the substance is present in the environment and in what concentrations?	Surveys and monitoring. Develop sensitive analytical methods. Apply analytical quality control procedures.
(b) How much of the substances enters the environment at what place, in what form and in what concentrations?	Determine quantities manufactured and used for different purposes, amounts disposed of as waste and eventual fate of material used in manufacturing. Identify any natural sources of the material.
(c) What is the fate of the substance?	Determine chemical and biological changes which the substance may undergo, the extent of its mobility in the environment.
2. Environmental effects	
(a) What, if any, effect is the substance observed to have on plant life?	Identify effects attributable to the substance. Determine circumstances and concentration of substance associated with the effects.
(b) What is the toxicity of the substance to plants and animals?	Determination of effects of substance on plants and animals under natural conditions. Investigation of effects of chemical state and other materials (soil, organic matter) on toxicity.
3. Remedial actions	
(a) What options are available for reducing or eliminating these effects?	Identify possible safer alternatives. Determine potential for reducing environmental input by improved technology. Develop alternative materials; develop cleaner technology. Estimate cost of applying each possible remedy.

20.2.1.2 *Nitrate*

Nitrate is synthesised naturally in the roots of plants and decomposed by bacteria in the presence of organic matter to nitrogen or ammonia. It is widely used as a fertiliser but much concern has been aroused by its presence in water, and its possible harmful effects on human health. Epidemiological studies in regions with high levels of nitrate in drinking water do not support the suggestion that the substance causes any increase in gastrointestinal cancer and methaemaglobinaemia is rarely diagnosed in developed countries. There is little or no evidence that nitrate causes significant environmental damage.

Methods are available for removing nitrate from water, but they add significantly to costs and it has been estimated that achieving a standard of 45 mg/l nitrate in potable water in the UK would add £16m per annum to costs. This expenditure should be balanced against the actual damage to health which can be shown to be attributable to nitrate in drinking water, and compared with the benefits to health which could accrue from alternative ways of spending the same sum of money.

It is probably not feasible to ban nitrogenous fertilisers but research is required on the form and timing of applications in order to optimise uptake by crops and minimise run-off and leaching to groundwater.

20.2.1.3 *Oxides of nitrogen and sulphur*

It cannot be denied that emission and deposition of sulphur and nitrogen oxides increases the acidity of poorly buffered waters. The situation is complicated by the effects of land usage in the catchment but acidification can have a serious effect on the ecology of fresh waters.

There are areas of the UK, noticeably in Wales, Scotland, and possibly the North West, which are affected but the scale of the problem is much smaller than in, for example, Scandinavia. As far as water quality is concerned, local remedial measures might prove effective at relatively low cost. A proper assessment of alternative remedies should, however, take account of the effects of acid deposition on other parts of the environment such as iron, stone work and forests. Additionally, although it cannot be proved, there is substantial evidence that UK emissions contribute significantly to the Scandinavian problem.

Some research is needed into the likely ecological consequences of different remedies, including reductions in sulphur emissions, to ensure that benefits are secured which bear some relation to the additional costs incurred.

The acid rain problem illustrates very well the difficulties of attributing a value to certain aspects of the environment when human health is not threatened, when the economic costs of pollution may not alone appear to justify the expenditure and when the environmental damage occurs across national boundaries.

20.2.1.4 *Organic compounds*

Pollution from organic substances requires research which is similar to that needed for metals. Added complications are the enormous diversity of substances from industry and from natural synthesis and transformation in the aquatic environment; thus biodegradation may eliminate them or convert them into more harmful substances and chlorination may give rise to chlorinated species with quite different properties. More information is needed on the nature of organic pollutants in the aquatic environment, the fate of harmful compounds and their transmission along the food chain.

20.2.2 Natural wastes

Natural wastes include domestic sewage and animal slurries. Sewage may be contaminated with industrial chemicals and animal slurries with agricultural chemicals, but their polluting effect otherwise arises from organic matter, whose biological destruction consumes oxygen, and ammonia. Domestic sewage is usually either treated to produce an effluent and a sludge, or discharged direct to coastal waters.

20.2.2.1 *Sewage effluent*

Sewage effluent is normally discharged to a water course. Its polluting load is destroyed naturally and, if the flow of the receiving waters is sufficient in relation to the discharge, no significant effects on the river can be detected. There are streams, however, where these conditions are not met and in extreme cases a substantial reach of the water course may become foul. In addition, many rivers receive storm waters and overflows which introduce a heavy polluting load over a short period of time.

The problems of chronic pollution of a water course are less technical than economic, but research is needed on the biological effects of chronic low levels of pollution and of the intermittent episodes of pollution, in order to provide objective criteria for upgrading or constructing treatment works.

20.2.2.2 *Sewage sludge*

Most sewage sludge is disposed of by spreading on agricultural land, by tipping on sacrificial land or by dumping at sea. Sewage which is collected from a mixed industrial and domestic population may be contaminated with metals and industrial chemicals. A large proportion of these are concentrated in the sewage sludge and become the limiting factor in the disposal options which are available.

It is likely to be many years before it is practicable to eliminate such contamination and it is essential, if we are not to incur large and unnecessary costs, that constraints on the disposal of sludge are based on a sound knowledge of the environmental effects of the contaminants rather than on emotional and subjective arguments. This knowledge would stem from the results of research on the environmental significance of chemicals, the need for which has been outlined above.

The chief problems in disposing of animal and human waste sludges are their volume, their offensive nature, and the presence of pathogenic organisms. It is possible by relatively simple treatment to reduce the numbers of pathogens in domestic sludge to acceptable levels and at the same time to reduce odour, but two of the main disposal routes, dumping at sea and spreading on agricultural land, are increasingly threatened with restrictions as a result of pressure from the European Community.

More research is needed to quantify, at least approximately, the extent of any damage caused to the marine environment by sludge disposal, together with any beneficial effects from the nutrients it affords. It is doubtful whether any additional effort in research into the agricultural benefits of sludge is justified.

20.2.2.3 *Coastal outfalls*

Some 15 per cent of sewage is discharged untreated through outfalls to coastal waters. Many of these were constructed many years ago to serve small communities where it would have been grossly uneconomic to instal treatment plant. Tourism and population growth have now in many places overtaken them, however, particularly where the outfall is short. Despite public concern the evidence is that the risk to health of people bathing in water close to a sea outfall is small in relation to other risks associated with bathing. Nevertheless it is acknowledged that beaches visibly contaminated with sewage solids are not acceptable and each of the regional water authorities concerned has a programme of expenditure to carry out improvements. The main limiting factor is money rather than technical knowledge.

Comparisons of the actual performance of new outfalls with the specification used in their design will show whether the mathematical models and data used in their design are satisfactory. This will indicate whether overdesign is resulting in excessive expenditure or, alternatively, whether the designs are failing to meet the intended standards on the beach.

There is a possibility that concern about the disposal of sewage sludge to sea will extend to the operation of coastal outfalls. Research is needed, therefore, to assess the environmental effects of sea outfalls so that decisions which may involve the expenditure of very large sums of money can be made on the basis of scientific fact rather than subjective opinion. Conventional treatment of the sewage inland will not only be substantially more expensive, but will result in the production of sewage sludge which will itself have to be disposed of. The coastal location means that only about half as much land is available for disposal as there could be at an inland site and this will increase the costs of transporting the sewage or sludge away from the coast. In addition, the effluent and its bacterial load will be discharged to a river and transported to the sea.

20.2.2.4 *Long-term needs – sewage treatment and sludge disposal*

Sewage represents a net cost to society in terms of its treatment and disposal.

Inevitably the cheapest possible solution for dealing with it will always be adopted. As a result of environmental concern and legislation, however, the costs of treatment and disposal will rise inexorably. Attempts have been made in the past to extract potentially valuable material from sewage or to find some means of converting it to a product with some intrinsic value. The digestion process generates methane which in some plants is collected and used as a source of energy but if the cost of collecting, storing and distributing the gas are taken into account, this is usually only economic for works serving populations in excess of about 100 000.

As treatment and disposal costs rise there is a need to examine the net cost of dealing with the sludge in different ways. Research into some alternative methods of treating sewage which produce less sludge, or allow separation of the heavy metals, should be increased even if their cost seems likely to be greater than that of conventional methods. Any new method will take decades to introduce and its economic attractiveness needs to be judged against the likely future costs of conventional technology rather than their current costs.

20.2.2.5 *Agricultural wastes*

Low cost methods of treatment for farm wastes are available although they are not always used. An increase in intensive animal husbandry creates environmental problems if insufficient attention is paid to environmental impact but the prevention of pollution from this source requires not more research but acceptance by the farmers of their responsibilities, and by consumers of the fact that environmental protection costs money. In the same way the use of herbicides and pesticides requires not more research but care and discipline in the way they are used.

20.2.3 Exploitation

20.2.3.1 *Farming and forestry*

The specific problem of animal slurry has been discussed above but farming and forestry have a more diffuse effect on the aquatic environment, especially close to a water course.

Eutrophication in reservoirs caused by phosphate entering from land run-off is difficult to deal with since very small quantities of phosphate are sufficient to permit excessive growth of algae. Various lines of research are being pursued, but none seems to offer particular promise. A cheap method of immobilising phosphate would be an attractive solution, but careful management in the application of phosphate fertiliser is also necessary.

Research into the effects of afforestation and de-afforestation on run-off water quality is in progress and this needs to be maintained and extended to an examination of how diffuse run-off from agricultural land contributes to river pollution.

20.2.3.2 *Angling*
Fish stocks in rivers frequently show marked variation. These fluctuations are attributed to a wide variety of causes from acid rain to over-fishing. A better understanding is needed of fish population dynamics in rivers and reservoirs so that natural variations do not cause unnecessary expenditure but that, on the other hand, if the change has a specific cause, it can be recognised and appropriate action taken.

20.2.3.3 *Land drainage, river improvements*
Major engineering works on the land adjacent to a river, or to the river channel itself, affect the ecology of the river and of the corridor of land through which it flows. There is scope for research to develop, if possible, engineering solutions which minimise damage to the river corridor. More urgent, however, is the need for some at least quasi-objective techniques for placing a value on a particular section of a river corridor so that attention is focused on areas of the highest environmental priority and potentially valuable schemes are not abandoned unnecessarily.

20.2.4 Aesthetic values
The weakest link in the chain of arguments for and against environmental protection is the value we place on the aesthetic aspects of the environment. For example, what is the commercial value of cleaning up a polluted beach where the sea is too cold for bathing? What would be the value of reducing the acidity of a lake whose acidity does not allow fish to survive?

One approach is to reduce everything to its readily calculable cash value, but when buying a car we do not base our choice on purely utilitarian calculations. In the same way we spend money on flowers which have no intrinsic value.

At present we have no objective means of measuring the value of aesthetic properties of the environment in terms which we can then use for comparison with the costs of achieving or maintaining them, and yet as private individuals, we constantly make similar judgements without much difficulty.

Perhaps one of the greatest challenges to research is, therefore, the development of satisfactory methods for placing a value on environmental protection in all its aspects so that priorities for spending can be properly allocated. Crude attempts have been made in the past but the subject deserves a much more concerted and determined attack.

20.3 CATEGORIES OF RESEARCH
Various types of research activity can be recognised and defined but in categorising research it must be borne in mind that the critical factor is not the activity but the motive of the scientist performing it. For the purposes of this chapter only two broad categories are considered, fundamental and applied.

(a) *Fundamental or speculative research*

This is carried out entirely for its scientific interest, without ulterior motive and with no practical end necessarily in view. It is undertaken primarily, but not exclusively, in the universities and research institutes. A basic principle is the freedom of the scientist to pursue his own line of research. It is a vital component of any research programme. In environmental research it may uncover previously unrecognised effects.

(b) *Applied research*

Research directed at finding a solution to, or achieving a better understanding of, a practical problem is applied research. It is sharply distinguished from fundamental research in that the provider of funds has the right, if he chooses to exercise it, to set practical objectives for the research.

(c) *Other types of research*

Reference is sometimes made to 'background' research and, for example, the Natural Environmental Research Council defines strategic research as 'bridging the gap between basic and applied research'. The acid test, however, is whether or not the work is being done with the aim, however distant, of achieving a better understanding of, or finding a solution to, a problem. If so it is applied research, otherwise it is basic. There is a risk that these and other categories are defined to attract support for research which is essentially basic by disguising it as applied.

20.4 FUNDING AND RESPONSIBILITY

Research needs have been identified on the basis of securing the best value for money in environmental protection for the nation. Responsibility for the applied research would fall partly upon government and partly on the manufacturing, agricultural, water and forestry industries.

In general, research aimed at securing the scientific understanding necessary for fair and sensible environmental legislation should be the responsibility of government; whereas research to ensure cost-effective compliance should be the responsibility of the polluting industries.

The Water Research Centre engages in extensive consultations with the water undertakings to determine their priorities for research and most other industries have comparable procedures.

The natural home for environmental responsibility is the Department of the Environment but, having gifted the Water Pollution Research Laboratory to WRC in 1974 and privatised the Hydraulics Research Station (now HR), it has no research laboratories of its own capable of carrying out research on the aquatic environment. Its needs are met through research contracts to, among others, NERC institutes, HR, universities and WRC.

The institutes comprising NERC were founded over a period of time to carry out fundamental research in various fields. Their funds were allocated as block grants from NERC which received the money from the Science Vote. The Rothschild principle was implemented to allow the 'customer' Government Departments a greater say in the direction of the Research Council programmes. Part of the Councils' funds are abstracted and given to the appropriate Government Department which returns them to the institutes in the form of research contracts.

The successful operation of the principle rests on four assumptions:

1. That the Department concerned has clear objectives and knows what it wants to do with the money.
2. That it has the resources to identify suitable contractors, to manage the contracts and to use the results.
3. That there will be no net reduction in the Research Councils' funds as a result.
4. That the institutes are willing and able to undertake the work required by the Department.

It is worth considering the validity of these assumptions in respect of DOE and NERC.

In the first place our parliamentary system does not encourage the long-term development of policy on which to base research objectives.

Secondly cutbacks in staff at DOE have undermined its technical and managerial resources.

Thirdly the institutes have failed to recover all the money abstracted from NERC under the Rothschild principle.

Finally the NERC institutes are not managed laboratories whose work can be quickly directed to new tasks. Traditionally the scientists enjoy a considerable measure of freedom in the research they choose to do and are not accustomed to accepting positive direction, although under the pressure of reductions in funding, attitudes are slowly changing.

A number of questions remain, therefore:

(a) Is sufficient money being spent on environmental research?
(b) If not where should the additional money come from and where should it be spent?
(c) If the funding is adequate is it being spent on the right problems?
(d) If not who should be responsible for redirecting it?
(e) What is an appropriate balance between fundamental and applied research?
(f) To what extent is it desirable that the universities and research institutes should be pushed into seeking contract research if this is detrimental to their basic research programmes?

(g) How can the responsible objective scientific views of environmental issues be presented to the public with an emotional appeal to match that of the extremist view?

There are no 'correct' answers to these questions but debating them might illuminate and improve decisions on expenditure and imbue those engaged in research with a real sense of purpose.

CHAPTER 21

Environmental protection – what can the nation afford?

LORD SHERFIELD
House of Lords Select Committee
on Science and Technology

21.1 INTRODUCTION

The House of Lords has a Select Committee on Science and Technology, as the House of Commons did until 1979, and I approach the subject of this conference as a member of this Committee. I have had the privilege of chairing three of our enquiries, into the scientific aspects of forestry, the provision of scientific advice to Government, and most recently the engineering aspects of the water industry.

The Select Committee is not another Royal Commission on Environmental Pollution, nor does it wish to be. Frequently, however, our enquiries take on an environmental flavour, and we find that environmental protection is one of the salient points that has to be considered. This arises for two main reasons. First, environmental protection is important, as the WRC's Conference indicates. Secondly, we have set ourselves, as a Committee, the task of investigating (*inter alia*) aspects of science and technology which are of public concern or where the interests of science and technology and the public *may* be at odds. Pollution of the environment has with great frequency been the outcome of technological advance and its side-effects. A new process, undertaken for laudable purposes, can improve human prosperity and yet cause damage to other parts of the environment.

The balance of advantage and disadvantage is a characteristic of all measures of environmental protection. Does the advantage of the polluting process outweigh the disadvantages? The difficulty of answering this question lies more often than not in the fact that like is not being compared with like. The advantages are frequently measured by criteria different from those of the disadvantages. In particular the advantages may be specific and localised whereas the disadvantages are more general and harder to assess; the advantages may be immediate and the disadvantages long term and uncertain. Where the former assist private profit and industrial prosperity, how are they to be balanced against the non-momentary considerations of quality of life, aesthetic appreciation of the environment and so on? When human advantage is set against disadvantage to animal, insect, plant and other communities, who adjudicates?

21.2 ASSESSMENT OF ADVANTAGE

The answers to the questions posed above, and the question which heads this chapter, are not static. They shift over time, and they vary according to the perspective of the person giving the answer. It would be unrealistic to expect immutable answers, and frequently it would be scientifically unsound as well. The balance shifts primarily according to:

(i) the appreciation of environmental threat;
(ii) the means available to reduce or eliminate that threat; and
(iii) the economic climate.

In a good economic climate the country may be willing to pay quite highly for environmental protection. In harder times, the cost of protection rises sharply when measured in lost jobs, lower production or more expensive goods and services. People prefer to live comfortably in poor surroundings than poorly in agreeable surroundings. But as an environmental threat becomes more obvious — either because scientific cause and effect become known or because the risk or rate of damage is recognised — the case for environmental protection grows and the perceived cost of protection falls. The assessment of threat and cost is also conditioned by personal priorities and understanding. According to the individual's standpoint, people judge situations very differently. For example, the assessments differ fundamentally in West Sedgemoor between the farmers and nature conservationists arguing about drainage, and at Sizewell B between the CEGB and the opponents of nuclear power. In each of these instances the arguments have tended to polarise around the party with a direct pecuniary interest in the alleged cause of environmental damage and the party with no pecuniary interest but a fear for long-term or even irreversible damage to the environment, human as well as natural. In extreme cases there are irreconcilable scientific judgements over which the parties disagree.

Such differences have to be resolved in a civilised society. This is achieved by political rather than scientific methods and by the decision of Parliament and Governments. Environmental legislation has been passed as a framework for the resolution of differences and to curb recognised excesses, acknowledging the need to protect the interest of the community at large from exploitation by the actions of a few and to protect the natural environment against short-term pollution in the long-term interests of future generations. Within this framework it is in principle possible to attempt to estimate the price of environmental

Ch. 21]   **Environmental Protection – What Can The Nation Afford?**   297

protection in each case and what the nation can, or should, afford. The Science and Technology Committee makes a contribution to this process by gathering the views of informed opinion on selected topics and weighing up the costs and benefits of scientific and technological developments, seen not only from the point of view of the developer but also from the wider perspective of public interest. This kind of political approach should characterise the actions of every senior administrator who has to take decisions concerning environmental protection.

21.3 PRINCIPLES OF ASSESSMENT

It is desirable that certain principles should guide political decisions about environmental protection and the following is a tentative list of such principles. It owes much to the Science and Technology Committee's experience but does not represent a formal Committee view.

21.3.1 Best environmental option

The British practice of relying as a rule on objectives for environmental protection rather than absolute standards has the merit of encouraging the adoption of new techniques for restraining pollution as soon as they become available and so achieving a consistently rising standard. This practice depends on effective control to ensure that the objectives are met by the best practicable means and that these do not degenerate into the cheapest tolerable means. The best practicable means, as enlarged by the Royal Commission on Environmental Pollution into the principle of the best practicable environmental option, should be adopted in any given circumstance. As the Government said in response to the Royal Commission, the logic of the best practicable environmental option is 'unassailable' and the option is a concept 'of considerable power and utility'. It should be adopted, with controls provided to ensure that it works.

21.3.2 The Polluter Pays Principle

Extreme bans on all activities capable of damaging the environment are not practicable. There are few black and white cases. Most human progress has been achieved at some environmental cost, and an environmental loss may well be compensated by an environmental gain. Environmental damage which is remediable is usually accepted by society, provided that the remedies are made; other activities which might be damaging can be rendered harmless by preventive action. In either circumstances, the prospective polluter should pay for the cost of the remedy. It is not acceptable that someone should make a profit at the expense of the public, by damaging the environment or imparing their enjoyment of the environment. The cost penalty of remedy should fall on the polluter and in turn his customers, not on the public at large. Not only is such a principle fair; it also creates an incentive to avoid pollution and favours processes which cause less environmental damage.

21.3.3 Irremediable damage

The polluter pays principle does not give anyone *carte blanche* to pollute even though financial recompense is offered. There should be a presumption against any form of irreparable damage, especially if an alternative is available, whether or not at higher cost.

21.3.4 Land use guidelines

Environmental damage can result from competition for scarce land resources. It is undesirable that land with a high value for one purpose should be used for another purpose for which the land's value is much lower. For instance land of exceptional nature conservation or agricultural value should not be taken for afforestation or reservoir construction, whereas lower quality land might be suitable. What is required is a policy of protection through land classification and guidelines on land use, so that land planning decisions are taken in full cognisance of the facts and priorities and with a hierarchy of land use values as a guide. This will require a new initiative by Government.

21.3.5 International co-operation

Commercial disadvantage is a powerful obstacle to environmental protection. What a nation can afford on its own is less than it can afford when its competitors adopt the same measures to restrict pollution simultaneously. Initiatives by the European Community and other international bodies in the field of environmental protection should therefore be welcomed and supported. The co-operative approach can surmount many commercial hindrances to good environmental practice.

21.3.6 Interest of the general public

The Government must accept the responsibility to speak for the interests of the general public of this and future generations in environmental matters, and to spend taxpayers' money on their behalf. As the Science and Technology Committee's enquiry into the potential of electric vehicles showed, there are occasions where a new development is in prospect with numerous advantages to the general public — in the case of electric vehicles primarily reductions in noise and air pollution — but the added costs of the development fall on the producer who does not stand to gain from them. It is proper in such circumstances for governments to subsidise development, in order to promote environmental benefits which might otherwise be stillborn.

21.3.7 Public opinion

Public opinion may at times be uninformed and emotive but it is powerful. Among the most important conditions for success in environmental matters are a healthy respect for public opinion and a recognition that public confidence is slow to build up but quick to dissipate.

In the Science and Technology Committee's enquiry into hazardous waste disposal, we pointed to the vital need for public confidence in the integrity and efficiency of the waste disposal industry. When disposal is conducted successfully the public do not ask for scientific justification for the disposal methods used; when a major accident has occurred the public become suspicious and will be extremely sceptical of every scientific justification, however accurate. Industry must always allow a margin of error in its processes and waste disposal, to ensure that serious incidents do not occur. The cost of mistakes, in the tighter controls and precautionary measures to restore public confidence, will outweigh the savings of any earlier parsimony. Responsible industries should be prepared to invest in environmental protection lest public confidence be lost — a case of 'keeping hold of nurse for fear of finding something worse'.

21.4 CONCLUSIONS

To lay down unexceptionable principles, set guidelines and prescribe good practice is straightforward; to apply them is complex. There are two main difficulties: the assessment of value and the assessment of risk. The assessment of value was referred to in the introduction to this chapter, and raises such questions as 'which has priority, the survival of an important industry or the survival of a species of watersnail?' The assessment of risk is complicated by the double standards associated with it. In older industries severe risks are willingly run and severe, sometimes devastating, consequences are cheerfully accepted. In newer industries, of which the nuclear industry is a specific example, there is, in some quarters, insistence on complete protection from risk where the risk is in practice negligible.

These assessments, which are in varying degree emotional, cannot satisfactorily be resolved by town meetings or public debate. In high technology these tend to be ill-informed. There will seldom be agreement on what is the best practicable environmental option in a difficult case. Such matters can be settled only by firm executive action of government. Yet governments, at least in Anglo-Saxon countries, tend to be anything but robust in facing and settling them.

Environmental protection should be based on a system of quality objectives, backed as appropriate by quality standards, with the standards gradually rising. The standards of one generation will not suffice for the next. Instead the objectives should respond to new technologies and in their turn act as a stimulus to the emergence of new technologies. The level at which the standards are set at any time is a matter of political judgement, taken in the light of prevailing circumstances. That judgement has to take account of economic and industrial considerations. It has also to reflect an obligation to future generations and to other forms of life within our environment. The reconciliation between the usually conflicting issues should in principle lead to the adoption of the best environmental option, and, again in principle, a country should not feel able to afford less than this.

Discussion

Chairman: **R. T. WHITELEY**
Chairman, Water Research Centre

Discussion opened with a comment on the general public's unwillingness to accept scientific judgements and evidence. In the opinion of one speaker this was due to the origins of the evidence. In many cases, as exemplified by the 'acid rain' debate, the research had been done at laboratories with a vested interest. Such links might prejudice the public's perception of the independent nature of this research and comment on these points was invited.

With regard to the independence of research, **Dr S. C. Warren** felt that if accusations were made of an industry that it was causing damage, such as the case of the CEGB with regard to acid rain, then it was inevitable that the industry would carry out research to investigate whether there was any evidence to substantiate such a claim. Thus it was unfair to discount such research based on the lack of independence. However, Dr Warren was aware that sometimes this research was done with little modesty and the results presented in a slanted way. Despite this there was no justification to doubt the objectivity of this research and Dr Warren thought that the CEGB could not be criticised for funding external research through the Royal Society in order to be seen pursuing an objective approach. Some comments followed on the problems of assessing costs and benefits in relation to environmental protection. Assessment of absolute costs was almost impossible and that probably of greater relevance was an assessment of changing costs and benefits. Assessment of the change in costs would require a great deal of effort on behalf of the authorities, but it could probably be done.

With regard to changes in benefit it is more difficult to obtain sensible results as assessment of the worth in improvement in aesthetic quality was fraught with problems. Similarly, with regard to more tangible benefits of environmental improvement, such as those associated with epidemiology and ecotoxicity research, it is very difficult to obtain definitive justification of the change in response in health terms or environmental terms that would accrue from a change in environmental management. It had been noted that several speakers had made the point that, in the end, any decision taken is made with greater regard for political priorities than for the scientific evidence presented. Thus, the authors were asked that, if this was the basis of the decision-making process, was scientific research either necessary or worthwhile?

Lord Sherfield responded by stating that every tool that was available to the decision maker should be used to provide information. These tools included cost—benefit analysis and scientific evidence. He thought it was a counsel of despair to state that one should not provide scientific evidence if the advice was to be ignored. Perseverance was required and Lord Sherfield believed that in some instances the scientific opinions proffered would influence a decision.

In answering the earlier point about the public's attitude to scientific and industrial opinion, **Lord Sherfield** noted that in the past, scientific opinion was accepted almost unquestioned by the public. However, as the speed of technological advance accelerated, the public at large became more confused and suspicious. This has led to a distrust of scientific, technical or 'informed' opinion. On the other hand the industrial world was seen to have a vested interest and thus their advice was also distrusted. Consequently, this distrust of scientific and industrial opinion has been exploited by some sectors of public opinion who are not interested in fact but rely more on their emotional instincts in reaching a decision in these matters. This exaggeration affects the political arena by making politicians less decisive for fear of appearing irresponsible which would lose votes. Hence governments tend to resort to delaying tactics, such as setting up enquiries and committees to study the problem in greater depth. These actions in Lord Sherfield's opinion lead to delay and were inefficient, but this was the reality of the situation. Consequently, he thought that there was great value in conferences such as this, which offered the opportunity of identifying ways of improving this decision-making process.

A research worker in industry revealed the dilemma facing a scientist. If he did not publish his work he was perceived as being secretive; conversely, publication of his work did not serve any purpose as this evidence was biased because of the vested interest. If one followed this logic then certain academic research workers in a given field also could be biased, because they had a vested interest in maintaining the momentum of environmental problems so that more research funds flow into their given area of study.

However, the point was made that all papers put forward for inclusion in the scientific literature were subject to review prior to publication. This was important as this process should prevent bias in the scientific literature.

Dr Warren continued this theme by observing that the water industry could be accused of being secretive because it was worried that it cannot prove that water was 100 per cent safe. However, it does have a lot of information which demonstrates that water does not cause harm and he thought the industry should make more capital out of this information. He continued by saying that because there was no medium to present the facts, the industry had succumbed to the temptation not to reveal the facts at its disposal and in his view this had created more problems than it had solved.

Discussion then turned to the problems of balancing benefits and costs of environmental protection. One delegate likened the environment to a huge capital asset and stated that much of the argument was based on how far one could go in terms of disinvesting in it. The benefits obtained in using the environment were usually short, whilst the costs in environmental terms are often very long. The point was also made that the environment is very diffuse; there was always a protagonist for any project, whilst there was seldom a focus for all the potential adverse effects. Thus the environmental lobby were thought to be disreputable on some occasions because they did not have a solid base for their arguments, whilst the protagonists did. This could be due to the fact that the environmental lobby was always trying to defend something about which it was very hard to be exact because of the diffuse nature of the environment. Politicians and research workers were urged to take more account of this diffuseness.

The problem of assessing the value of environmental protection was again referred to. Delegates seemed to be agreed that there is no absolute way of doing this. Comparisons may need to be made in terms of time, but comparisons also need to be made in terms of space. To illustrate this point, the vast amount of money spent in the UK on research into the problems caused by high nitrate levels in drinking water was cited. During the last twenty years, ten cases of methaemoglobinaemia had been reported. This expenditure was then compared with that of other countries and particularly third world countries, where 25 000 people a day were dying from the lack of a safe water supply. Thus it could be concluded that there were other ways in which we could compare our performance in the field of environmental protection with that of other countries and gain some perspective as a consequence.

Lord Sherfield's point regarding public disillusionment with scientific opinion was returned to. Several examples of changes in medical practices associated with the use of radiations such as X-rays, which had been discontinued because of the discovery of the detrimental health effects of some types of radiation were quoted. The speaker continued by stating the difficulties in communicating the balance between the risk of the potential detrimental effects from certain medical practices and the benefits which could also be obtained. These kinds of risks, as well as those associated with the risk of death from following various occupations, had been examined in a report of the Royal Society. The difficulties were increased because mistakes made in the past by scientists, (as exemplified by the dangerous medical practices now forsaken), reduced the credibility of the scientific community to make authoritative statements. In this speaker's opinion the risks associated with detrimental environmental effects are so small that the occupational risks associated with

Discussion

research into the environment, involving data collection for example, were much greater.

Mr G. Lean took up the comments on costs and the amount the public is prepared to pay for environmental protection. In Germany the public were apparently quite willing to pay large amounts in relation to the problems caused by 'acid rain' because the damage to forests in that country, associated with this problem, seemed to be large. Mr Lean thought that this expenditure by the notably reactionary Christian Democrat government was due to the pressure of public opinion, most notably the green movement.

Another speaker took up this last part of the discussion. He noted that a recent opinion poll in Germany had posed the question, 'Would you be willing to see electricity prices increase by 15% in order to deal with the acid rain problem?' The anser was an overwhelming 'yes' and this fact should be recognised as an example of public commitment to environmental protection.

The speaker continued by stating that many environmental issues have a large emotional content and thus to informed opinion public opinion will continue to appear to be irrational. This was part of the price of living in a democracy, and commentators on environmental matters, including journalists, scientists and industrial representatives, have a responsibility in seeking to inform public opinion. If in trying to do this commentators are too ardent, then this will only serve to discredit their views. Thus the experts must try harder to communicate their views to the general public.

Summing up

PROFESSOR R. W. EDWARDS
Deputy Chairman, Welsh Water Authority

There seem to be two basic strategies in summing-up conferences. The first is to write the summing-up before the conference — sometimes even before receiving the papers. This approach, in my view, induces undue anxiety and confusion in the delegates who, learning about issues for the very first time at the end of the conference, think they've dozed too frequently in conference sessions.

I must admit to adopting a rather different approach, of purging my mind of prejudices to make room for others to be drawn in, of scribbling comments throughout the conference and of burning the midnight oil on the last, and for most delegates, particularly jovial evening, seeking, in vain, to put ideas into some logical order. How so much easier for the paper author who can slip off after his contribution for a relaxing gin. Unfortunately, adopting my approach, I couldn't do that and during these last three days I've felt a little like Bessie Braddock, that formidable Labour MP who apocryphally, having told Winston Churchill that he was drunk, suffered the riposte; I know Bessie and *you* are ugly — and what is more, I shall be sober in the morning! The trouble with adopting this rather passive, responsive approach is that I shall say nothing new — merely acting as a very ineffective and personal sieve of what you have heard before.

I should like to start with Lord Ashby's excellent opening address and with the question he posed 'Why protect the environment?'. This same theme was reiterated by Lord Sherfield in Chapter 21. As Lord Ashby pointed out, in the past the answer has been a very **anthropocentric** one, that is we protect the environment to protect ourselves. But as Lord Ashby indicated, we are now becoming far more **biocentric** in our outlook (my jargon — not his), that is we are showing concern for other species (particularly when they are rare) from which man derives no direct, demonstrable benefit. Even the anthropocentric element has extended its territory both in space and time: Lord Sherfield, in Chapter 21, refers to this obligation to future generations. In Wales this obligation shows prosaic but tangible expression in the way the water authority apportions

Summing Up 305

its fishery costs — one-third of which is allocated to the conservation of fish stocks for present and future generations to appreciate and is charged on the environmental services account without any perceptible whimper.

But Ron Packham gave a very timely reminder (Chapter 7) that our spatial consciousness needs to go much further, when referring to the sharp contrast between expenditure in the UK over a few possible cases of methaemoglobinaemia resulting from increased nitrate concentrations in water and the enormous numbers of deaths in the third world from waterborne diseases, where we were told very early in the confidence that $10 to $15 was all that was required to save a life. Anthropocentricity clearly needs a wider horizon.

It is important that we consider how far a biocentric ethic might develop because it has a profound effect on environmental standards and associated costs. Now it is rare species and maybe biological diversity which are of concern. Although we've learned to accept that some ecological change is a consequence of the discharge of organic and other wastes to the environment, tomorrow the maintenance of all species naturally occurring within a river may become a legitimate objective of pollution control — or even the protection not only of populations but of individual organisms from pollution-induced mortality. You might think my scenario is ridiculous but who would have thought a few decades ago that a rare fish, the snail darter, would have delayed, for several years, the completion of a major dam construction project in the USA.

If we shift towards this biocentric ethic we shall become increasingly concerned with protecting the physical as well as the chemical environment of aquatic organisms. I raise the matter of the physical environment because, although not the subject of the conference, modifications through land-use changes and by river-engineering, have a major impact on what lives in rivers and other surface waters and these considerations are very germane to the question of environmental protection.

I should now like to comment on the discussion about the role of toxicology in establishing quality standards. With man there are many problems, not least of which are that we prefer not to conduct tests on ourselves, that we are not concerned only with the average man but also with individuals in the most vulnerable groups, and that exposure and damage may be separated by a substantial time interval. Inevitably, judgements about standards are based on data from several sources and involve the application of 'safety factors'. Sir Frederick Warner's chapter (Chapter 3) put the risks of pollutants in water supplies comfortably into perspective, a perspective reinforced by the case-studies described in other chapters.

Dr Patricia Fraser (Chapter 5) demonstrated for us some of the shortcomings of epidemiological studies but despite their essentially correlative approach and their retrospective nature, they sometimes provide a reassuring safety-net when establishing safety standards for drinking water. Nevertheless David Young posed the question about the wisdom of providing alternative water-supplies at Worksop — a decision taken by the Severn-Trent Water Authority some while ago — following earlier studies which suggested an implied NO_3—cancer link. 'Knowing what we know now would we have invested that money?'. My answer would be that we don't live in a perfect world or have a limitless supply of flawless crystal balls and, with the evidence then available, the decision was sensibly cautious and the approach one we would all wish managers to take.

With aquatic organisms, the problems are in some respects different — experiments can be conducted but the trouble in the past has been that these experiments have rarely been ecologically relevant in mimicking responses under natural conditions. Nevertheless with fish, despite these theoretical criticisms, great progress towards realistic standards has been made from simple laboratory studies of dose—response relationships. In this particular field WRC has been pre-eminent.

I should like to make a general comment about safety factors — which are in great part expressions of ignorance or uncertainty. I was mildly perturbed to hear of the apparent equanimity with which one speaker applied safety factors of 1, 2 or even 3 orders of magnitude — the cost to industry of such safety factors would be enormous and it would clearly be sensible in most cases to invest in further research to reduce this degree of uncertainty to more acceptable 'intrinsic' limits.

I now wish to come to the important issue of establishing compliance with standards from sampling programmes — for the practical constraints of sampling are frequently inadequately considered when standards are established. I was delighted with many conclusions in the chapter by Mr Ellis and Dr Miller (Chapter 8), not least their concern to stress the importance of trying to find out how the system, which one intends to sample, behaves, as a prerequisite for efficiency in programme design. It is a lesson I've tried to preach to scientists of the Welsh Water Authority, with only modest success. There is a strong tendency in the water industry to collect data at great expense and then to ignore it when its analysis would lead to an understanding and predictive capability as well as a reduction in sampling needs and costs. But water authorities are not the only guilty parties in this context, Central Government is equally guilty — it continues to brood over the water quality archive and, since the demise of the Water Data Unit, little of substance has hatched from the costly accumulations of tapes and floppy discs.

I cannot avoid making some comment about discussions at this Conference on international obligations and particulary those relating to the attitude of the European Commission with respect to environmental protection. Despite the assurances that participants received from Mr Fairclough (Chapter 10) about the good intentions of the Commission I detected a residual anxiety. In my view there are several practical steps the Commission might take to improve its credibility. The Commission is primarily responsible for the first published draft of all Directives and at this stage — before the Directives have reached the

Summing Up 307

political arena with its element of bargaining — the Community deserves draft instruments with far more scientifically acceptable (and defensible) standards. The Commission after all has free choice to seek the best technical and scientific advice and it should seek it more assiduously. The House of Lords Select Committee has also expressed the view that the internal scientific expertise in the Environmental Directorate of the Commission needs further strengthening — a view I very much support.

Inevitably the advantages and disadvantages of the environmental quality standard (EQS) and uniform emission standard (UES) approaches have been given a thorough airing. Whilst there is a logic about the EQS, problems arise in implementation through effluent control — particularly where reductions in emissions prove necessary. Water authorities seem to have some distance to go before they can do this with justifiable assurance.

The advantage with the UES is that once you've accepted its dubious logic, administrative implementation is easy. Our own practical conversion to EQSs, although sensible if unduly strident, is relatively recent and it is not surprising that our EEC partners view this conversion with some scepticism. We have also wandered from the path of 'best practical (or technical) means' only recently — again for very good reason as Dr Farquhar so ably indicated (Chapter 18). This path, even with List 1 substances, can be seen as ludicrous when taken beyond the point where significant benefit — even when temporally and spatially stretched — is achieved for further cost. I must admit, however, that the discussion on this issue of the difference between UK and Community viewpoints, was conducted by the principals, Mr Gunn and Mr Fairclough, in the best possible taste and it reminded me of the story of the Bishop and Admiral who had rather disapproved of each other's attitudes for a long time and who met on Victoria Station. Having been to a Palace Garden Party they were dressed in full uniform and robes. The Bishop approached the Admiral and said, 'Porter, is this the right platform for Guildford'? 'Oh yes, Madam', replied the Admiral, 'but do you think you should be travelling in your condition'?

I must now come to research, referred to in Dr Warren's chapter (Chapter 20), and particularly the changes in funding over the past few years with its possible consequences.

I shall comment, at the beginning, that like many others, I cannot accept the sharp, distinctive, binary approach of Dr Warren of academic research on the one hand and applied research on the other — this is far too simple. There is a whole complex of motives, activities and time scales for research which leaves a middle ground of strategic research where the approach may seem academic but where the motive and direction are applied even if the end is not sharply focused towards a product.

The Rothschild arrangement, which transferred funds from Research Councils, such as NERC, to Government Departments for commissioning specific programmes of research in the Research Councils, has brought about a major shift from strategic research areas, for which they were intended, to applied and generally short-term studies. In WRC a similar shift towards clearly identifiable applied topics has occurred — in this case because of a change in management attitudes and particularly their responsiveness to their paymasters, principally the water authorities, in shaping research programmes. Some of you may regard these shifts in emphasis as wholly admirable but in my judgement there are substantial dangers in retracting from the middle ground of strategic research. Furthermore, these shifts have also occurred at a time of a general withdrawal of research funds by Central Government through DOE in the field of water research.

I was delighted that, following the debate of its Report on the Water Industry, Lord Sherfield's Select Committee on Science and Technology received an assurance by a Government Minister that a Research Requirements Committee would be set up to examine the consequences of the depletion of central research funding. I hope too that it, or some other influential group, such as the Royal Commission on Environmental Pollution, will find the time to take case-studies like 'acid rain', where had we carried out appropriate research and maybe looked for environmental signals more carefully we might have anticipated and avoided some of the current and impending environmental damage. That more than a decade after the phenomenon and its effects on aquatic ecosystems was identified in Scandinavia, we are caught out by discovering acid streams in parts of the UK with severely depleted fish stocks, seems remarkable and hardly to the credit of our environmental research and surveillance systems.

There are many other issues at this Conference which, in contrast, gave me comfort, in particular the efficient way some water authorities, like Wessex, have developed a code of practice for spillages and dangerous discharges (Chapter 14). I look forward very much to the findings of the proposed WHO 'epidemiological' investigation with respect to such accidents reported by Dr Gilad (Chapter 15). In Wales most of these accidental discharges now emanate from farms — and about 80% of all prosecutions for pollution are from this one sector. In the UK, changes in farming practice, particularly with animal units, require a more thoughtful and effective response then they are receiving at the moment.

My time is up but I cannot resist ending with one of the gems which surfaced at this Conference — the dilemma posed by Dr Farquhar in his chapter (Chapter 18) of choosing to eat roast beef on the banks of a smelly river or to eat sausage on the banks of a non-smelly one. Despite my professional allegiance to the pollution control lobby, one of long-standing, it brought home to me the stark realisation of how highly I regard food. Thank goodness I have never been faced with this stark choice!

APPENDIX I

Support paper
The perception of risk

P. POWELL and R. F. LACEY
Water Research Centre

1. INTRODUCTION

Warner [1] has described the types of pollution that can occur in drinking water supplies, has indicated the sizes of the risks that are involved, and has compared them objectively with other risks encountered in everyday life. It would be rational to conclude that such information, together with knowledge of the costs of risk abatement, should be sufficient for decisions to be taken on how such risks should be managed. There are, however, additional factors that are not wholly dependent upon the estimated sizes of the risks, but which strongly affect how different kinds of risks are *perceived* by the public. It is the perceived risk that usually determines the public's reaction to a suspected hazard, and the extent to which pressure is brought to bear on policy makers, via one or other of the channels that are available in a democratic society.

Discussion of the perception of risk is not new and has been well covered by the Royal Society Study Group [2]. The present note aims to highlight the main factors that influence the perception of risk and to discuss how they may operate in the context of the water industry. The risks that we have in mind are those that involve specific damage to human life or health, rather than more general damage to the environment, although the same ideas may be developed to consider these issues too.

2. FACTORS AFFECTING THE PERCEPTION OF RISK
2.1 Involuntariness of exposure

Risks associated with voluntary activities, such as smoking or hang gliding, are more readily accepted than risks to which people are exposed involuntarily, such as carcinogens in food. Since in general people have no choice over their water supplies (other than by buying bottled water, or installing point-of-use treatment), risks via the water supplies might be expected to give rise to more concern than a voluntary risk of similar magnitude.

2.2 Concentration of events

A motorway pile-up, or a lifeboat disaster attracts more attention than the same number of deaths occurring on the roads or at sea in ones and twos over an extended period. In our own industry, the parallel might be the cholera outbreak as opposed to the increased mortality from heart disease in soft water areas. The diffuse nature of deaths from the chronic diseases tends to lessen the concern that would otherwise be felt for a similar number of deaths occurring all at once.

2.3 Immediacy of consequences

A risk with a delayed outcome can be more acceptable than a risk of similar probability where the outcome is immediate. Smoking would appear in the first category; crossing the road in the second.

Since any possible influence of water supplies on the chronic diseases will be delayed-in-effect, this factor will operate to lessen concern.

2.4 Unfamiliarity of hazard

New, unfamiliar, or 'unnatural' hazards (for example from new food additives, radiation from nuclear power) are regarded as worse than risks from established, familiar or natural causes, such as traditional foods, cosmic radiation, or radioactive emissions from Aberdeen granite.

This factor probably helps to account for the lack of concern over the water softness/CVD story — nothing is being done to the water, it's just the way the water is. Conversely, the risk of cancer, through adding fluoride to the water, is not acceptable to some people, even though there is little evidence of an effect.

This factor may operate differently, depending upon which cause-and-effect we are considering. For example, any suggestion that asbestos from AC pipes is having a health effect might generate concern because AC pipes would probably be considered 'new' and 'unnatural'. Lead from plumbing ought to come into the same category, but seems not to, maybe because of its longer history of use. If upland or lowland surface supplies were found to be contributing to raised cancer rates, it might depend on what was causing the effect. If it were a natural component of such waters, there might be less concern than if the culprit were man-made, such as a compound present in industrial effluents.

2.5 Risk without benefit

Where non-beneficiaries of a particular action take the risks, then such risks are considered less acceptable than cases where the beneficiaries themselves are at risk. An example of this is the vaccination against whooping cough of older children (who would normally survive an attack, but who are at slight risk from the vaccination process) for the benefit, by preventing epidemics, of young children.

This factor presumably operates in the fluoride argument. The beneficiaries are the children, or at least those with their own teeth, whereas the supposed cancer risk is to adults, although once fluoridation has been established for a generation or so then the same children who have benefited become the adults at risk.

Apart from this example, those who have the benefits of a water supply are usually also those at risk from any potential hazard it might bring. This factor would not therefore be expected to operate adversely in our field.

2.6 Imposition on society

Where the action of an individual increases the risk to himself, rather than to others or society as a whole, then such risks may be considered acceptable. This factor operates favourably in the case of smoking, although this may change if the dangers of passive, or second-hand, smoking are proven.

The sort of risks that we are considering are risks to society, rather than to the water industry itself, so this factor should operate to increase concern over such risks.

A closely related factor to this one is the question of who pays to reduce a certain risk. If it is the individual who is at risk who has to pay, this is more acceptable than if someone else has to foot the bill. There again, with most water-related risks, the individual at risk would not be expected to pay for any remedial action.

An example of a water-related problem in which this factor may be operating to lessen concern is the question of lead in water. Where it is that part of the service pipe belonging to the consumer which is causing the problem, it is the individual with the lead pipe who is at risk, and he may, if he wishes, pay to put it right. Some responsibility for remedial action, however, has been accepted by the water industry and government. This may tend to heighten the perception of risk, as if it had been imposed.

2.7 Named individual put at risk

Considerable sums on money are spent on air/sea rescue, for example, when a named individual is at risk. Similar sums spent in reducing more diffuse risks may be more cost effective, but do not have the same public appeal. This factor would be expected to make funding less easy, (that is make the risk more acceptable) in the cases we are considering.

2.8 Secrecy

Risks arising from secret activities, in the defence field, for example, are regarded as worse than those derived from open activities.

Assuming that water supply continues to be regarded as an open activity, we can expect that this factor should not give rise to increased concern.

2.9 Possible partiality of the risk evaluator

Risks evaluated by groups who are suspected of partiality are perceived as worse than risks evaluated by neutral groups. An example of this would be statements by the nuclear power industry about the safety of its own installations.

From this point of view, pronouncements by the water industry about its own risks (or by the WRC, on behalf of the industry) might be regarded less favourably than conclusions from a body that appeared to be 'independent'.

2.10 Multiple factors

It should be pointed out that in most cases several of the factors described above may be operating at the same time, and many are often associated with each other. For example, there are many instances of voluntary activities whose risks are immediate-in-effect, to the individual himself, who has to pay to ensure his own safety. A major part of the work by psychologists and social scientists in this field has been to attempt to develop methods for identifying and separating the effects of such factors.

3. PARTICULAR WATER ISSUES

Let us consider how the factors described in section 2 might affect the perception of risk of four types of health effect that are currently in some way linked, rightly or wrongly, with water quality.

The first type of effect is the possible relationship between a chronic disease (for example, cardiovascular disease) and a natural feature of the water supply, such as hardness.

Secondly there is the question of adding fluoride to water supplies to improve dental health.

Thirdly there is the health effects of lead in relation to the problem of plumbo-solvency. Finally there is the question of whether bathing on 'polluted' beaches causes gastro-intestinal upsets. For comparisons with these four 'water' issues we also consider how risk perception affects proposals to build nuclear power stations.

If we consider each of these in turn against the factors outlined in section 2, we can construct the table as shown in Table 1. In this table a star (★) indicates that the factor is operating to increase concern, a dash (−) indicates the factor minimises concern, and some cases, shown by a question mark (?), are either debatable, or it could depend on how the situation is handled. The totals on the line give the number of stars in each column.

Notice that the more stars appear under a given effect, the greater appears to be the public concern over the possible risks. Nuclear power (6★) is debated

Appendix I – Support Paper

Table 1

Factor	CVD and hardness	Lead	Bathing beaches	Fluoride	Nuclear power
Involuntariness	★	★	–	★	★
Concentration	–	–	?	–	★
Immediacy	–	–	★	–	?
Unfamiliarity	–	?	?	★	★
Risk without benefit	–	–	?	★	★
Imposition of society	?	?	★	★	★
Named individuals	–	?	?	–	?
Secrecy	?	?	?	?	?
Partiality of evaluator	?	?	?	?	★
	1	1	2	4	6

in Parliament, features in political manifestos, and public enquiries are usually set up before such power stations can be built. Fluoridation (4★) has given rise to pressure groups and court cases. On the other hand, little attention appears to be paid by the public and press to water lead, and still less to the 'water story' for CVD, each of which only receives one star in Table 1.

4. DISCUSSION

What conclusions can be drawn from the kind of analysis outlined above? We would like to suggest that there are at least three ways in which the water industry, by intelligent self-help, can mollify rather than heighten the perception of those risks that the industry is responsible for managing.

The first way is concerned with openness of information. An appearance of secrecy or remoteness on the part of the industry (or government department?) would tend to heighten suspicion that there may be something to worry about. Frankness – especially in situations where we genuinely do not know – is ultimately beneficial, even though it may be less comfortable in the very short term.

The second way in which we can help ourselves concerns the independence of evaluation. The industry should welcome evaluations of health hazards by outside groups, provided that they are properly informed of the facts and are aware of the intricacies of many water supplies. This is an argument for some diversification of health-related research, a situation which presently obtains.

Thirdly, the industry should try to avoid making any future hazard seem as 'new', unfamiliar or unnatural to the general public. This may be an argument for releasing information bit by bit, as it becomes available, rather than saving up the whole story for a possible bombshell. This argument reinforces the first issue, that of openness of information.

By taking care over such factors such as these, the water industry may avoid awarding a given risk a few extra 'stars' that could turn a relatively minor effect into the *'raison d'etre'* for a new pressure group.

5. CONCLUSIONS

There are factors other than the numerical size of a risk that affect the degree of concern with which a given risk is perceived by the general public. Such factors apply to risks for which the water industry is responsible.

Some of these factors, mainly those involving public relations, can be controlled by the industry, which therefore has an influence on whether present or future problems become major public issues.

6. REFERENCES

[1] Warner, Sir Frederick (1983) Pollution — the risk in perspective. Chapter 3 of this book.
[2] Warner, Sir Frederick (ed.) (1983) *Risk Assessment*. The Royal Society.

List of Delegates

AGG, A. R., Manager, Marine Pollution Control, WRC Environment.
ALABASTER, Dr J. S., Consultant, Pollution and Fisheries, 1 Granby Road, Stevenage, Hertfordshire.
AL-AWADI, Dr F., Deputy Director, Water Resources Centre, Environment Protection Council, PO Box 24395, Safat, Kuwait.
ALEXANDER, B., Director of Planning, North West Water, Dawson House, Great Sankey, Warrington, Cheshire.
ASHBY, FRS, THE LORD, The House of Lords, Westminster, London.
BAKER, P. J., Group Leader, Greater London Council, DG/AE/TRG Branch, County Hall, London.
BARNDEN, A. D., Senior Quality Officer (Rivers and Groundwater), Anglian Water, Ambury Road, Huntingdon, Cambridgeshire.
BARNHOORN, H. T., Chief Scientist, Eastbourne Waterworks Company, 14 Upperton Road, Eastbourne, East Sussex.
BASKETTER, F. B., North West Water, Dawson House, Great Sankey, Warrington, Cheshire.
BATES, Dr A. J., Scientist, Anglian Water, Lincoln Division, Waterside House, Waterside North, Lincoln.
BAYES, C., Scientist, WRC Scottish Office, 1 Snowdon Place, Stirling, Scotland.
BAXTER, K., Research Student, University of Manchester Institute of Science and Technology, Department of Civil and Structural Engineering, Sackville Street, Manchester.
BAXTER, K. M., Scientist, WRC Environment.
BEK, P., Metropolitan Water, Sewerage and Drainage Board, PO Box A53, Sydney South, Australia.
BELL, M. G. W., Scientist, Anglian Water, Norwich Division, Yare House, 62–64 Thorpe Road, Norwich.
BELLAK, J. G., Chairman, Severn-Trent Water, Abelson House, 2297 Coventry Road, Sheldon, Birmingham.
BIRD, Dr P., Divisional Scientist, Anglian Water, Oundle Division, 1 North Street, Oundle, Peterborough, Cambridgeshire.

BJERRE, F., Deputy Secretary, Oslo and Paris Commission, New Court, 48 Carey Street, London.
BOLAS, P. M., Chief Scientist, Mid Kent Water Company, High Street, Snodland, Kent.
BOWDEN, K., Quality Planner, Thames Water, Nugent House, Vastern Road, Reading, Berkshire.
BREEMEN, Dr L. W. C. A. VAN, Chief, Biology Department, Water Storage Corporation, PO Box 61, 4250 DB Werkendam, The Netherlands.
BROOKS, D. R., Principal Pollution Control Officer, Severn-Trent Water, Leicester Water Centre, Gorse Hill, Anstey, Leicester.
BROWN, J. C., Deputy Product Safety Manager, The Clayton Aniline Company, PO Box 2, Clayton, Manchester.
BUCHANAN, D., Director and River Inspector, Highland River Purification Board, Strathpeffer Road, Dingwall, Ross and Cromarty, Scotland.
BURFIELD, I., Deputy to Water Quality Controller, Essex Water Company, Langford, Maldon Essex.
BUSTARRET, M., Ingenieur Principal, Compagnie Générale des Eaux, 52 Rue D'Anjou, 75008 Paris, France.
CALCUTT, T., Manager, Sludge Group, WRC Processes.
CAMPBELL, S. C., Supply and Treatment Engineer, Bristol Waterworks Company, PO Box 218, Bridgwater Road, Bristol, Avon.
CATTELL, Dr F., Consultant to Metropolitan Water Sewerage and Drainage Board, Macquarie University, North Ryde, Sydney, Australia.
CHAVE, P. A., Principal Scientific Officer, South West Water, Matford Lane, Exeter, Devon.
CHILDS, J., Pollution Control Officer, Severn-Trent Water, Mapperley Hall, Lucknow Avenue, Nottingham.
CLAYTON, R. C., Director, WRC Processes.
COLE, J. A., Manger, Analysis and Monitoring, WRC Environment.
COLLEY, M. D., Principal Pollution Control Officer, Northumbrian Water, Northumbria House, Regent Centre, Newcastle upon Tyne, Tyne and Wear.
COLLINGE, V. K., Director, Planning, Water Research Centre.
CRABTREE, Dr R. W., Research Fellow, Department of Civil Engineering, University of Birmingham, Birmingham.
CRAENENBROECK, Dr W. VAN, First Chemist, IV Antwerpse Waterwerken NV, Mechelsesteenweg 64, 2018 Antwerpen, Belgium.
CREWE, S., Divisional Water Quality Planner, Wessex Water, PO Box 9, King Square, Bridgwater, Somerset.
CROWTHER, J. H., Environmental Adviser, BP Chemicals Limited, 76 Buckingham Palace Road, London.
DAVIES, A. S., Executive Officer, Strathclyde Regional Council, Strathclyde House, 20 India Street, Glasgow.

List of Delegates

DAVIES, J. N., Senior Engineer, Aspinwall and Company, 5 Swan Hill Court, Shrewsbury, Shropshire.

DENNER, J., Environmental Engineer, Balfours, Yeoman House, 63 Croydon Road, London.

DIGNUM, D., Principal Scientist, Ardleigh Reservoir Committee, Clover Way, Ardleigh, Colchester, Essex.

DOBBS, Dr A. J., Section Head, Environment Section, Princes Risborough Laboratory (BRE/DOE), Princes Risborough, Aylesbury, Buckinghamshire.

DUNN, P. J., Pollution Control Officer, British Airports Authority, Head Office, Gatwick Airport, West Sussex.

EDGINGTON, P. G., Consultant, BP Oil International Limited, Britannic House, Moor Lane, London.

EDWARDS, P. R., Assistant Environment Adviser, ICI Mond Division, PO Box 8, The Heath, Runcorn, Cheshire.

EDWARDS, Professor R. W., Head, Department of Applied Biology, University of Wales Institute of Science and Technology, King Edward VII Avenue, Cardiff, Wales, and Vice-Chairman of Welsh Water Authority.

ELLERKER, R., Scientist, Operations, Anglian Water, Oundle Division, North Street, Oundle, Peterborough, Cambridgeshire.

ELLIS, J., Head, Statistics, WRC Environment.

ELMS, N., Research Student, 7 The Bourne, Fleet, Hampshire.

EVANS, Dr G. P., Head, Intake Protection, WRC Environment.

EVESON, J., Chief Scientific Officer, Department of the Environment (NI), Water Services Headquarters, Parliament Buildings, Stormont, Belfast, Northern Ireland.

FAIRCLOUGH, A. J., Director for the Environment, Commission of the European Communities, Rue de la Loi 200, B-1049 Brussels, Belgium.

FARQUHAR, Dr J. T., Group Environmental and Protection Manager, Albright and Wilson Limited, 1 Knightsbridge Green, London.

FARRIMOND, Dr M., Principal Planner, Severn-Trent Water, Abelson House, 2297 Coventry Road, Sheldon, Birmingham B26 3PR.

FAWELL, J., Head, Toxicology, WRC Environment.

FERGUSON, R., Assistant Director Finance, North West Water, Dawson, House, Great Sankey, Warrington, WA5 3LW.

FIELDING, M., Head, Organics, WRC Environment.

FISH, OBE, H., Chief Executive, Thames Water, New River Head, 177 Rosebery Avenue, London.

FLOYD, W., W. D. Floyd Water Researches, 41 Glycena Road, London.

FORD, G. S., Scientific Officer, Southern Water, Guildbourne House, Worthing, Sussex.

FRASER, Dr P., Department of Medical Statistics and Epidemiology, London School of Hygiene and Tropical Medicine, Keppel Street, London.

FRENCH, V. H., Editor, Water Services, Fuel and Metallurgical Journals Ltd, Queensway House, 2 Queensway, Redhill, Surrey.

GARDINER, E. R., Quality Controller, North Surrey Water Company, The Causeway, Staines, Middlesex.

GARLAND, J. H. N., Manager, Freshwater Pollution Control, WRC Environment.

GASCOINE, I. S., Chief Water Quality Officer, Southern Water, Kent Division, Capstone Road, Chatham, Kent.

GILAD, Dr A., Consultant, Environmental Systems Management, World Health Organization Regional Office for Europe, 8 Scherfigsvej, 2100 Copenhagen, Denmark.

GOFTON, B. F., Division Environment Adviser, ICI Petrochemicals and Plastics Division, PO Box 90, Wilton, Middlesborough, Cleveland.

GRAY, Dr N. F., Lecturer in Environmental Sciences, Trinity College, University of Dublin, Dublin 2, Eire.

GRIFFITHS, P., North West Water, Cheshire Effluent Treatment and Supply, Weaverham Grange, Hertford, Northwich, Cheshire.

GUNN, J. A. L., Under-Secretary, Department of the Environment, Water Directorate, 43 Marsham Street, London.

HALE, J., Principal, Department of the Environment, Room B456, Romney House, London.

HAMER, A. D., Water Quality Planning Officer, Southern Water, Hampshire Division, Eastleigh House, Market Street, Eastleigh, Hampshire.

HAMMERTON, D., Director, Clyde River Purification Board, Rivers House, Murray Road, East Kilbride, Glasgow.

HARDWICK, Dr D. C., Pollution Scientist, Ministry of Agriculture Fisheries and Food, Great Westminster House, Horseferry Road, London.

HARPER, E., Chief Scientific Adviser, North West Water, Dawson House, Great Sankey, Warrington, Cheshire.

HARPER, W. R., Director of Finance, Thames Water, New River Head, 177 Rosebery Avenue, London.

HARRISON, Dr D., Vice-Chancellor, Keele University, Keele Park, Keele, Staffordshire.

HARVEY, R., Principal Pollution Control Officer, Severn-Trent Water, Shelton, Shrewsbury, Shropshire.

HAWES, F. B., Biological Adviser, Planning, Central Electricity Generating Board, Laud House, 20 Newgate Street, London.

HEATHER, D. J., Company Secretary, Fertiliser Manufacturers Association, Greenhill House, 90–93 Cowcross Street, London.

HELM, M., Water Quality Officer, Northumbrian Water, Northumbria House, Regent Centre, Gosforth, Newcastle upon Tyne, Tyne and Wear.

HENDERSON, B., Chairman, Water Research Centre and Chairman, Anglian Water, Ambury Road, Huntingdon, Cambridgeshire.

List of Delegates

HILL, J. A., Senior Engineer, Severn-Trent Water, Abelson House, 2297 Coventry Road, Sheldon, Birmingham.

HOWE, M. C., District Water Quality Officer, Anglian Water, Great Ouse House, Clarendon Road, Cambridge.

HOWELLS, V., Divisional Scientist, Severn-Trent Water, Lower Severn Division, Southwick Park, Gloucester Road, Tewkesbury, Gloucester.

HUGGINS, R. J., Wessex Water, Avon and Dorset Division, 2 Nuffield Road, Poole, Dorset.

HUGHES, L. B., Principal Technical Officer, North West Water, Rivers Division, PO Box 12, Newtown House, Buttermarket Street, Warrington, Cheshire.

HUNT, Dr D. T. E., Head, Analysis, WRC Environment.

HYDES, O., Principal Pollution Technical Officer, Department of the Environment, Room B456, Romney House, London.

ISAAC, Professor P. C. G., Chairman, Civil Engineering Planning Panel, Science and Engineering Research Council, Polaris House, North Star Avenue, Swindon, Wiltshire.

IQBAL, J., Laboratory Manager, PPC Consultants Limited, Crown House, Copthorne Bank, Crawley, Sussex.

JAPPINEN, H. J., Senior Chemist and Biologist, Jaakko Poyry Consulting OY, PO Box 16, 00401 Helsinki, Finland.

JENKINS, W. R., Head, Fish Section, WRC Environment.

JOHNSON, D., Biologist, Intake Protection, WRC Environment.

JONES, A. N., Divisional Scientist, Welsh Water, Gwynedd Division, Penrhosgarnedd, Bangor, Gwynedd.

JONES, D. GRUFFYDD., Director, Central Directorate on Environmental Pollution, Department of the Environment, Romney House, 43 Marsham Street, London.

JONES, F., Principal Scientist, North West Water, Dawson House, Great Sankey, Warrington, Cheshire.

JORDAN, A. C., Commercial Operations, Water Research Centre.

KEY, A. R., Principal Pollution Control Officer, Severn-Trent Water, Tame Division, Tame House, 156–170 Newhall Street, Birmingham.

KIFF, Dr R. J., Lecturer, Pollution Research Unit, University of Manchester Institute of Science and Technology, PO Box 88, Manchester.

KING, Dr N. J., Head, Toxic Substances Division, Department of the Environment, Room A3.335, Romney House, 43 Marsham Street, London.

KINNERSLEY, D., Adviser on Water Institutions, 111 Church Street, Chesham, Buckinghamshire.

KNAPP, A. B., Environmental Co-ordinator, IMI plc, PO Box 216, Witton, Birmingham.

KONEMANN, Dr W. H., Ministry of Housing, Physical Planning and Environment, Postbus 439, 2260 AK Leidschendam, The Netherlands.

List of Delegates

KRUIJF, Dr H. de, Head, Chemical Biological Division, National Institute for Water Supply, PO Box 150, 2260 AD Leidschendam, The Netherlands.

LACEY, R. F., Head, Epidemiology, WRC Environment.

LACK, Dr T. J., Operational Studies, WRC Environment.

LAING, J., Nutrition Chemical Department, Ministry of Agriculture, Fisheries and Food, Woodphorne, Wolverhampton, West Midlands.

LEAN, G., Environment Correspondent, *The Observer*, 8 St Andrews Hill, London.

LEATHERLAND, T. M., Estuary Survey Officer, Forth River Purification Board, Colinton Dell House, West Mill Road, Colinton, Edinburgh.

LEE, R., Recovery Support Officer, Wessex Water, 303 Preston Road, Yeovil, Somerset.

LUCAS, J. L., Technical Manager, Aspinwall and Company, 5 Swan Hill Court, Shrewsbury, Shropshire.

McLARTY, R. M., Advisory Microbiologist, School of Agriculture, Kings Buildings, West Mains Road, Edinburgh.

MADDOX, J., Editor, *Nature*, 4 Little Essex Street, London.

MAIDMENT, C., Manager, Commercial Operations, WRC Environment.

MANCE, Dr G., Head, Environmental Quality Objectives, WRC Environment.

MARIS, P. J., Head, Landfill Section, WRC Environment.

MARSH, R. J., Senior Consultant, Aspinwall and Company, 5 Swan Hill Court, Shrewsbury, Shropshire.

MARTIN, D. E., Environmental Scientist, BP International Limited, Britannic House, Moor Lane, London.

MASON, C., Divisional Technical Officer, Anglian Water, Lincoln Division, Waterside House, Waterside North, Lincoln.

MASSCHELEIN, W. J., Director, CIBE, Waterloosesteenweg 764, B-1180 Brussels, Belgium.

MAYER, M., Journalist, Environmental Data Services, Finsbury Business Centre, 40 Bowling Green Lane, London.

MESS, H., Post Graduate Research, Technology Policy Unit, Aston University, Gosta Green, Birmingham.

MILLER, D. F., Deputy Director, Tay River Purification Board, 3 South Street, Perth, Scotland.

MILLER, Dr D. G., Assistant Director, Environmental Contamination, WRC Environment.

MILLS, Dr A. R., Environment Officer, Welsh Water, Wye Division, St Nicholas House, St Nicholas Street, Hereford, Wales.

MODHA, P. M., Deputy Secretary, Government of Gujarat Health and Family Welfare Department Sachivalaya, Gandhinagar 382010, India.

MONK, Dr D. C., Co-ordinator, Ecotoxicology, British Petroleum, Environmental Control Centre, Britannic House, Moor Lane, London.

List of Delegates

MOORE, A. E., Assistant Co-ordinator, South West Water, Matford Lane, Exeter, Devon.

MORTIMER, Dr D. J., States Chemist and Analyst, States of Guernsey Water Board, St Saviours Laboratory, The Reservoir, Guernsey.

MOSS, Dr J., Manager, Sludge Disposal and Operational Studies, WRC Environment.

NEWMAN, L. E., Technical Information Manager, Water Research Centre.

NEWMAN, Dr P. J., Head, International Office, Water Research Centre.

NOORDAM, Dr P. C., Toxicologist, KIWA, PO Box 70, 2280 AB Rijswijk, The Netherlands.

O'CONNOR, C. N., Senior Engineer, Waterworks, Dublin Corporation, Castle Street, Dublin 2, Eire.

ODELL, R. I., Press and Publications Manager, Water Research Centre.

O'DONNELL, T., Environmental Quality, WRC Environment.

PACKHAM, Dr R. F., Assistant Director, Water Contaminants, WRC Environment.

PAINTER, Dr H. A., Head, Biodegradability, WRC Environment.

PALFRAMAN, J. F., Senior Scientific Officer, Laboratory of the Government Chemist, Cornwall House, Stamford Street, London.

PALMER, D. J., Divisional Supply Scientist, Wessex Water, Bristol Avon Division, Quay House, The Ambury, Bath, Avon.

PARKINSON, A., Senior Scientific Officer, North West Water Rivers Division, PO Box 12, New Town House, Warrington, Cheshire.

PEARCE, A. S., Principal Pollution Technical Officer, Department of the Environment, Room B456, Romney House, 43 Marsham Street, London.

PECKHAM, D., Administration Scientist, Thames Water, Eastern Division, The Grange, Crossbrook Street, Waltham Cross, Hertfordshire.

POWELL, P., Epidemiology, WRC Environment.

PRICE, D. R. H., Scientist, Anglian Water, Cambridge Division, Great Ouse House, Clarendon Road, Cambridge.

RAHEEM, M. Y. A., Superintendent, Pollution Control Division, Environment Protection Council, PO Box 24395, Safat, Kuwait.

RAYNER, C. S., Research Engineer, Esso Petroleum Company Limited, Esso Research Centre, Abingdon, Oxon.

RICHARDS, W. N., Assistant Director, Strathclyde Regional Council, Water Department, 419 Balmore Road, Glasgow.

RIDGWAY, Dr J. W., Manager, Water Quality and Health, WRC Environment.

RIPTON, T. S., Principal Quality Officer, Anglian Water, Rivers House, Springfield Road, Chelmsford, Essex.

RODDA, D. W. C., Head, Water Research Management, Department of the Environment, Water Directorate, Room B4/53, Romney House, 43 Marsham Street, London.

RODDA, Dr J. C., Natural Environment Research Council, Institute of Hydrology, Maclean Building, Crowmarsh Gifford, Wallingford, Oxon.
ROLLEY, H. L. J., Head, Pollutant Chemistry, WRC Environment.
ROUSE, M. J., Director, WRC Engineering.
RUSSELL, Dr P. W., Research Scientist, British Gas, London Research Station, Michael Road, London.
SALEM, Dr A. Y., Professor of Chemical Engineering, University of Alexandria, Alexandria, Egypt.
SALZWEDEL, Professor J., Chairman, The Council of Experts on Environmental Matters, Lennestrasse 35, D-5300 Bonn 1, Federal Republic of Germany.
SANDLAND, R., Quality and Safety Manager, Borden UK Limited, North Baddesley, Southampton, Hampshire.
SCHIECKE, U., Federal Agency of the Environment, Bismark Platz No. 1, D-1000, West Berlin 33, Federal Republic of Germany.
SCOTT, R. N., Divisional Scientist, Welsh Water, West Wales Division, Meyler House, St Thomas Green, Haverfordwest, Dyfed, Wales.
SCUPHOLME, P. L., Head of Environmental Services, BP Petroleum Development (NW Europe), Farburn Estate, Dyce, Aberdeen, Scotland.
SELBY, K., Principal Pollution Control, Severn-Trent Water, Derwent Division, Raynesway, Derby.
SEMPLE, A. G., Secretary, Water Authorities Association, 1 Queen Anne's Gate, London.
SHERFIELD, THE LORD, House of Lords Select Committee on Science and Technology, Westminster, London.
SMALLS, I. C., Senior Biologist, Chemical Sub-Branch, Metropolitan Water, Sewerage and Drainage Board, PO Box A53, Sydney South, Australia.
SMITH, A. H., Divisional Trade Effluent Officer, North West Water, Merton House, Stanley Road, Bootle, Merseyside.
SNOEK, O., Head of the Laboratory Department, Municipal Water Works Amsterdam, Leidsewaartweg 73, 2106 NB Vogelenzang, The Netherlands.
SOLBE, J. F., Manager, Fish, Toxicity and Biodegradability, WRC Environment.
SOLMAN, A., Manager, Instrumentaiton, WRC Environment.
SOUTHWOOD, Professor T. R. E., FRS, Chairman, Royal Commission on Environmental Pollution, Church House, Great Smith Street, London.
SPEARS, A. T., Deputy County Engineer, Donegal County Council, County House, Lifford, Eire.
SWANWICK, K. H., Resources and Treatment Manager, Yorkshire Water, Western Division, Broadacre House, Vicar Lane, Bradford.
TARBOX, M., Divisional Water Quality Manager, Wessex Water, PO Box 9, King Square, Bridgwater, Somerset.
TAYLOR, D., Analytical and Information Services Manger, ICI Brixham Laboratory, Brixham, Devon.

List of Delegates

TAYLOR, L. E., Chief Engineer and Head, Water and Environment Protection Division, Welsh Office, Cathays Park 2, Cardiff, Wales.

THOMPSON, A. J., Assistant Chemist, Sunderland and South Shields Water Company, 29 John Street, Sunderland.

THORBURN, J., Assistant Chief Engineer, Scottish Development Department, Pentland House, 47 Robb's Loan, Edinburgh.

THORNE, D., Laboratory Manager, Yorkshire Water, North and East Division, Tosti, 20 Avenue Road, Scarborough, Yorkshire.

TIDY, G. L., Company Pollution Prevention Manager, The Boots Company plc, 1 Thane Road, Nottingham.

TOMS, R. G., Chief Scientific Officer, Wessex Water, Wessex House, Passage Street, Bristol, Avon.

TYRRELL, V., Economic Adviser, Department of the Environment, Romney House, 43 Marsham Street, London.

URQUHART, Dr C., Principal Scientific Officer, Yorkshire Water, West Riding House, Albion Street, Leeds, Yorkshire.

WALKER, D., Divisional Manager, Severn-Trent Water, Upper Severn Division.

WARD, G., Tutor, Hatfield Polytechnic, Hatfield, Hertfordshire.

WARD, T., Division Pollution Prevention Officer, Yorkshire Water, PO Box 201, Broadacre House, Vicar Lane, Bradford, West Yorkshire.

WARING, Dr M., Medical Officer, Department of Health and Social Security, Hannibal House, Elephant and Castle, London.

WARN, Dr A. E., Senior Planner, Systems, Anglian Water, Ambury Road, Huntingdon, Cambridgeshire.

WARNER, Sir F., FRS, Emeritus Partner, Cremer and Warner, Consulting Engineers, 22 Beach Road, Hartford, Northwich, Cheshire.

WARREN, Dr S. C., Director, WRC Environment.

WATSON, C., Divisional Scientific Officer, Yorkshire Water, Castle Market Building, Exchange Street, Sheffield.

WEBB, L. J., Group Head, Environmental Research, PIRA, Randalls Road, Leatherhead, Surrey.

WHITAKER, R. J., Regional Emergency and Security Officer, Wessex Water, Wessex House, Passage Street, Bristol, Avon.

WHITE, R. J., Assistant Secretary, Water Authorities Association, 1 Queen Anne's Gate, London.

WHITELEY, R. T., Member, Thames Water Authority.

WHITLEY, Dr R. J., Divisional Water Quality Planning Officer, Wessex Water, Avon and Dorset Division, 2 Nuffield Road, Poole, Dorset.

WIDDOWS, Dr J., Institute for Marine Environmental Research, Prospect Place, The Hoe, Plymouth, Devon.

WILLETTS, D. G., Divisional Scientist, Severn-Trent Water, Trinity Square, Horninglow Street, Burton upon Trent, Staffordshire.

WILSON, Dr K. W., Senior Scientist, North West Water, Dawson House, Great Sankey, Warrington, Cheshire.

WOOD, P. C., Officer in Charge, Ministry of Agriculture Fisheries and Food, Fisheries Laboratory, Burnham on Crouch, Essex.

WOOD, T. B., Chief Chemist, The Colne Valley Water Company, Blackwell House, Aldenham Road, Watford, Hertfordshire.

YOUNG, D. D., Assistant Director, Severn-Trent Water, Abelson House, Coventry Road, Sheldon, Birmingham.

YOUNG, J. A., Director of Operations, Wessex Water, Wessex House, Passage Street, Bristol, Avon.

YOUNG, S. N., Post-Doctoral Research Fellow, University College of Wales, Penglais, Aberyswyth, Dyfed, Wales.

ZERIBI, T., Sanitary Engineer, Adviser on Environmental Hazards and Pollution Control, World Health Organization Regional Office of the Eastern Mediterranean, PO Box 1517 Alexandria, Egypt.

Index

Acceptable daily intakes, 80
Accepted engineering practice, 170, 171, 173, 177, 178
Acid,
  budget, 34
  mine drainage, 34
  rain, 27, 31, 34, 275, 287
  wastes, 148
Afforestation (*see* Forests)
Aldehydes, 42, 45
Alkanes, 45
Alkylaromatics, 45
Alpha-BHC, 48
Aluminium, 34, 35, 39
Ammonia, 35, 241
Analytical quality control (AQC), 50
Annex I substances, 144
Annex II substances, 145
Annex III, 145
Anoxia, 40
Anthropogenic, 33, 40, 41, 43, 45
Antimony, 35
Aquatic herbicides, 41
Aquatic toxicology, 94
Arsenic, 35, 65, 151
Asbestos, 65, 310

Bathing, 289, 312
Bathing beaches, 86, 289, 312
Bathing water, 162
Beaches, 289, 312
Benzene, 46
Bioaccumulation, 145
Biochemical oxygen demand, 30
Biodegradability, 145
Black list, 150, 174, 187
Boron, 35, 64
Bottom deposits (*see* Sediments)
Bromine, 35

Cadmium (*see also* Heavy metals), 35, 39, 41, 144, 151, 154, 155, 187
  compounds, 151
Calcium, 66
Cancer, 64, 277, 287, 310
  stomach, 89
Capital expenditure, 259, 271
Carbohydrates, 43
Carbon dioxide, 277
Carbon tetrachloride, 41
Carboxylic acids, 44
Carcinogen, 65, 77
Cardiovascular diseases, 87, 312
Chemical accident, 219, 228
Chemical constituents, 110
Chemical control, 163
Chemical emergencies, 220
Chemicals, industrial, 285, 288
Chemical industry, 267, 269
Chemical stress, 100
Chlorinated hydrocarbons, 65
Chlorine, 241
Chloroform, 41, 44, 47
Cholera, 85
Chromium, 35, 39, 151
Coastal outfalls, 289
Coastal waters, 38, 40
Cobalt, 35, 39
Codes of practice, 143, 271
Community environmental policy, 158
Community's Environmental Action Programme, 158
Compliance, 123
Conservation, 261, 263
Control of Pollution Act 1974, 30, 183, 199
Copper (*see also* Heavy metals), 35, 39, 41, 151
Corn, 276
COST 64B, 41

Cost benefit, 264, 276
Cost effectiveness, 262
Cost of protection, 296
Cost of sampling, 133
Costs, 133, 259, 260, 261, 263, 264, 275, 276, 278
Cupric, cuprous (*see* Copper)
cyanide
Cyanide, 41

DDT, 48, 49
Definition of pollution, 144
Deforestation (*see* Forestry)
Department of the Environment, 292
Detergents, 57
Dieldrin, 48
DIN standards, 177
Directive 76/464/EEC, 186
Directive on the quality of water for human consumption, 109
Directives, 162, 186, 263
Discharge
 consents, 145, 171, 205
 direct, 30, 185
 indirect, 170
 individual, 184
 municipal, 170, 176
 standards, 153, 170, 184, 185, 186, 187
Drinking water, 39, 42, 162
 pollutants, 44
 standards, 108, 109

EC, 109, 181, 262
 Council of Ministers, 160
 directives, 33, 35, 98, 119, 133, 161, 162, 186
 duties, 152
 Environmental Action Programme, 186
 maximum admissible concentration, 248
 commitment, 181, 186
Ecotoxicology, 94
Effluent standards (*see* Discharge standards)
 tax, 170, 171, 174, 175
Emergency exercises, 216
Emergency plans, 215
Emergency response systems, 226
Environmental control, 268, 272
Environmental cost, 270
Environmental damage, 297
Environmental impact assessment, 164
Environmental legislation, 142, 265, 278, 296
Environmental option, 299
Environmental pressure groups, 134

Environmental problems, 268, 296
Environmental protection, 198
Environmental Protection Agency, 111, 121
Environmental quality, 253, 260, 264, 265
Environmental Quality Objectives, 153, 156, 161, 181, 187, 188, 299
Environmental Quality Standards, 153
Environmental threat, 296
Environmentally clean, 271
Epidemiology, 56
Esters, 44
European Community (*see* EC), 109 121, 158, 186, 271, 283, 298
Eutrophication, 28, 35, 40, 290
Expenditure, 261, 269
 by industry, 269

Federal Emission Protection Law, 172
Ferric, ferrous (*see* Iron)
Fertiliser, 287
Field drainage (*see* Land drainage)
Fishing, 291
Fluoridation, 90, 312, 313
Fluoride, 311, 312, 313
Food stress, 100
Forestry, 290
Forests, 28, 34, 166
Freshwater fish, 162

Gamma-BHC, 48
Gastro-intestinal disease, 86, 312
Global Environment Monitoring System, 162
Grey list, 151, 187
Groundwater, 31

Halogenated organic compounds (*see* Organohalogens)
Hardness (total) (*see* Water hardness)
Hazchem Code, 210, 211
Hazfile, 211
Health, 56, 268
Health and Welfare Canada, 111, 121
Health effect, 277, 312
Heart disease, 66
Heavy metals, 285, 288, 290
Herbicide, 65,
Hodgkin's disease, 277
Hormones, 67,
Hydrocarbons,
 (aromatic) (*see* Polyaromatic hydrocarbons)
 (halogenated) (*see* Organohalogens compounds)

Index

Incineration at sea, 147
Industrial water supply, 169
Industry, 269
   sewage treatment, 178
   water supply, 169
Insecticide, 48, 65
Iodine, 35
Iron, 35, 39

Ketones, 42

Land drainage, 291
Land use guidelines, 298
Leachates, 27
Lead, 35, 39, 67, 151, 274, 277, 310, 311, 312, 313
   in petrol, 166, 274
List I substances, 150, 161
List II substances, 161, 187

Magnesium, 67
Maize virus, 276
Mammalian toxicology, 73
Manganese (*see also* Heavy metals), 35, 39
Marine, 42, 214
Marine environment, 142, 289
Marine water, 42, 46
Maximum Admissible Concentration, 128
Mercury, 35, 39, 40, 144, 151, 155, 187
   compounds, 151
   from industry, 153
Metalloids, 35, 36, 38
Metals, 285, 288
Methaemoglobinaemia, 35, 64, 88, 287
Micronutrients, 35
Mobile Intake Protection System, 251
Molybdenum, 35, 39
Monitoring, 123, 130
   marine environment, 155
Multiple determinands, 129
Multispecies tests, 99
Mutagens, 65

N-nitroso, 64, 89
Nickel, 35, 39, 41, 151
Nitrogen, 39
   N-nitroso, 64, 89
   nitrate, 39, 64, 88, 89, 241, 287
   nitrite, 65
   nitrogen, (ammoniacal) (*see* ammonia)
   oxides, 166, 287
No-effect level, 77, 95

Odour, 241
Oral contraceptives, 67

Organic carbon, 42
   compounds, 287
Organic compounds,
   of phosphorus, 151
Organic matter, 42
Organic pollutants, 41, 72
Organic substrates, 42
Organic sulphides, 45
Organohalogens, 144, 147, 151
Organoleptic substances, 151
Organosilicon, 144
Oslo and Paris Commissions, 142
Oslo convention, 143, 148

Paraffins (*see* Hydrocarbons)
Paris Convention, 143, 149, 150, 151
Pathogens, 67
Pesticides, 48, 147
Petrol, 250, 274
Pharmacokinetics, 78
Phosphorus, 35, 151
   phosphate, 35, 57, 64
Phytotoxicity (*see* Toxicity)
Pipelines, 150
Pollution, 46, 199, 204, 205, 295
   air, 150, 166, 268
   by dumping, 143
   control, 36, 161, 163, 165, 173, 181, 199, 200, 201, 202, 203, 205, 233, 249, 250, 275
   definition, 144
   detection, 241, 250
   diffuse, 163
   elimination, 151
   groundwater, 163
   incidents, 239
   inorganic, 34
   land-based sources, 149
   natural, 27
   sea by hydrocarbons, 163
   thermal, 27
Polyaromatic hydrocarbons, 46
Polychlorinated biphenyl (PCB), 47
Polychlorinated biphenyls, 48, 154, 155
Polynuclear aromatic hydrocarbons, 46
Polysaccharides, 43
Prevention, 163, 165
Prevention of Oil Pollution Act 1971, 203
Prevention of pollution, 159
Principle
   anticipation, 170, 172
   polluter pays, 159, 181, 185, 269, 297
   proportionality, 174
   Rothschild, 293
   unified management, 181
   assessment, 297

Index

Prior Consultation Procedure (PCP), 145, 147
Probability distribution, 137
Proportionality principle, 174
Proteins, 43
Public, 134, 185
Public concern, 279, 295, 297, 298
Public debate, 299
Public opinion, 279, 298
Public water supply, 169

Random sampling error, 136
Research
    needs, 81, 283, 286
    policy, 283
    strategic, 292
    technical, 149
Reservoirs, 290
Revenue expenditure, 260
Risk perception, 312
Risks, 55, 309, 310
Rivers, 291
    basin, 182
    corridor, 291
    intakes for potable supply, 239
    pollution, 45, 198
    quality, 39, 124, 184
Rivers (Prevention of Pollution) Acts, 182, 203

Sampling
    constraints, 129
    frequency, 131
Sediments, 46
Selenium, 35
    Engineering Association, 177
    disposal (*see* Sludge disposal)
Sewage effluent, 43, 288
Sewage Engineering Association, 177
Sewage disposal (*see* Sludge disposal)
Sewage sludge, 288
Sewage treatment, 289
Sewerage, 182
Sewers, 268
Shellfish, 162
Silage, 243
Silicon, 151
Silver, 35, 39
Sludge, 30
    disposal, 289
Special Effluent Act, 169
Speciation, 99
Specific ion electrode, 242, 249
Speculative research, 292

Spillages,
    inland, 210, 213
    oil, 250
    oils chemicals, 72, 209
Standard deviation, 136
Standards, 81, 123, 128, 252, 263, 285
    interpreting, 171
    percentile, 128, 137, 139
Storm, sewage overflows, 200
Storm waters, 288
Strippable Pollutant Monitor, 250, 252
Strontium, 35
Subsurface drains (*see* Land drainage)
Sulphide, 40
Sulphur, 40, 166
    oxides, 268, 275, 287
Surface water, 162

Tap water, (*see* Drinking Water)
Tastes and odours, 34, 241
Thermal pollution, 27
Tin, 35, 151
TiO_2, 149
Titanium, 35
    dioxide, 148, 149, 162
Toluene, 46
Toxicity, 36
    acute, 73, 97
    chronic, 73, 97
    of mixtures, 102
Toxicokinetics, 78
Toxicology, 72
Trace constituents, 144
Trace elements, 35, 36, 39, 40
Trace metalloids, 35
Trace metals, 35, 36
Trihalomethanes, 65
Type I error (alpha), 136
Type II error (beta), 136
Types of limit, 114

Unauthorised pollution of water, 171
Underground dereliction, 204
Unified management, 182, 188
Uniform emission standards, 153, 156
United Nations Environment Programme, 162
United States Public Health Service, 109, 111, 112
Urea, 42
Uronic acids, 42
US Environmental Protection Agency, 112

Vanadium, 35, 39
Ventilation frequency, 244

Index

Viruses, 67
Volcanic eruptions, 28, 31
Voluntary risk, 309

Waste agriculture, 290
Waste animal, 289
Waste disposal, 143, 147, 148, 299
Waste gas scrubbing, 173
Waste industrial, 169, 180, 200
Waste law, 170
Waste natural, 288
Waste producer, 148
Water, 42
   cycle, 182
   hardness, 87
Water Acts, 169, 170, 171, 173, 179 183
Water Incident Officer, 215
   quality, 119
   supply, 182, 268
   pollution (*see* Pollution control)
   table (*see* Groundwater)
   treatment (*see* Drinking water)
Water Research Centre, 101, 292
   Mk III fish monitor, 244
Water-carriage system, 268
Whaling, 279
Whole cells pollution monitors enzymes 248
Wildlife, 279
Wildlife species, 277
Wood inquiry, 275
Working groups, 176
World Health Organization, 109, 111, 112
   guidelines, 114

Xylene, 46

Zinc (*see also* Heavy metals), 35, 41, 151